Photothermal Science and Techniques

PHYSICS AND ITS APPLICATIONS

This series of short texts on advanced topics for students, scientists and engineers will appeal to readers seeking to broaden their knowledge of the physics underlying modern technology.

Each text provides a concise review of the fundamental physics and current developments in the area, with references to treatises and the primary literature to facilitate further study. Additionally texts providing a core course in physics are included to form a ready reference collection.

The rapid pace of technological change today is based on the most recent scientific advances. This series is, therefore, particularly suitable for those engaged in research and development, who frequently require a rapid summary of another topic in physics or a new application of physical principles in their work. Many of the texts will also be suitable for final year undergraduate and postgraduate courses.

1. **Electrons in Metals and Semiconductors**
   R.G. Chambers
2. **Basic Digital Electronics**
   J.A. Strong
3. **AC and DC Network Theory**
   Anthony J. Pointon and Harry M. Howarth
4. **Nuclear and Particle Physics**
   R.J. Blin-Stoyle
5. **Thermal Physics**
   Second edition
   C.B.P. Finn
6. **Vacuum Physics and Techniques**
   T.A. Delchar
7. **Basic Electromagnetism**
   E.R. Dobbs
8. **Quantum Mechanics**
   Second edition
   Paul C.W. Davies and David S. Betts
9. **Statistical Physics**
   Second edition
   Tony Guénault
10. **Photothermal Science and Techniques**
    D.P. Almond and P.M. Patel

Photothermal Science and Techniques

DARRYL ALMOND

Reader in Materials Science, University of Bath, UK

and

PRAVIN PATEL

Photothermal Consultant

CHAPMAN & HALL

London · Glasgow · Weinheim · New York · Tokyo · Melbourne · Madras

Published by Chapman & Hall, 2–6 Boundary Row, London SE1 8HN, UK

Chapman & Hall, 2–6 Boundary Row, London SE1 8HN, UK

Blackie Academic & Professional, Wester Cleddens Road, Bishopbriggs, Glasgow G64 2NZ, UK

Chapman & Hall GmbH, Pappelallee 3, 69469 Weinheim, Germany

Chapman & Hall USA, 115 Fifth Avenue, New York, NY 10003, USA

Chapman & Hall Japan, ITP Japan, Kyowa Building, 3F, 2–2–1 Hirakawacho, Chiyoda-ku, Tokyo 102, Japan

Chapman & Hall Australia, 102 Dodds Street, South Melbourne, Victoria 3205, Australia

Chapman & Hall India, R. Seshadri, 32 Second Main Road, CIT East, Madras 600 035, India

---

First edition 1996

Typeset in Times 10/12 by Florencetype Ltd, Stoodleigh, Devon
Printed in Great Britain by St Edmundsbury Press,
Bury St Edmunds, Suffolk

ISBN 0 412 57880 8

A catalogue record for this book is available from the British Library
Library of Congress Catalog Card Number: 95–71226

♾ Printed on permanent acid-free text paper, manufactured in accordance with ANSI/NISO Z39.48-1992 and ANSI/NISO Z39.48-1984 (Permanence of Paper).

Contents

Preface

The purpose of this book is to give a concise introduction to the physical principles utilized in photothermal science and to explain its novel techniques and applications. Photothermal science is a rapidly developing area of research that has produced a number of powerful analytical instruments that are in widespread usage in both the academic and commercial sectors. This book is directed primarily at the technologists, research workers and postgraduate students who are new to this scientific area, who require a comprehensive summary of this area of physical science. A thermal wave treatment has been employed throughout as the adoption of the familiar techniques of wave physics helps in the understanding of many novel photothermal phenomena.

Extensive use has been made of the ideas and discoveries of numerous workers in this field without the referencing that would be found in a primary text. At the end of each chapter there are Reference/Further reading sections containing references to some of the many texts used by the authors in preparation of this work.

The authors are indebted to the numerous workers that have granted permission for the reproduction of their figures to illustrate parts of the text of this book. Full references to the sources of these figures are in the References at the end of each chapter.

Symbols and abbreviations

A	Thermal wave amplitude at point of generation
a	Defect radius
α	Thermal diffusivity
α_t	Linear thermal expansion coefficient
b	Ratio of effusivities in media 1 and 2
β	Optical absorption coefficient
C, c	Specific heat
c	Velocity of light in vacuum
γ	Ratio of specific heats at constant pressure and constant volume
D	Semiconductor carrier diffusivity
D_n, D_t	Normal and transverse OBD (mirage) deflections
e	Electron charge
e_a	Effusivity of medium a
E_g	Energy gap
$E(z)$	Electric field component of an electromagnetic wave at point z
ε	Emissivity
ε_0	Permittivity of free space
$\varepsilon^*(\omega)$	Complex dielectric constant at angular frequency ω
f	Frequency
ϕ	Thermal wave phase angle
Γ	Effective thermal wave reflection coefficient
h	Convective heat transfer coefficient
$h\nu$	Quantized photon energy
I_0	Optical intensity
j	$\sqrt{-1}$
k	Thermal conductivity
L	Coating or layer thickness

λ	Wavelength
m	Electron mass
N^*	Complex refractive index
N_f	Free electron carrier concentration
$N(z, t)$	Carrier concentration at z, t
n'	Real part of refractive index N^*
n''	Optical attenuation coefficient, imaginary part of N^*
η	Non-radiative recombination efficiency
P_0	Ambient pressure
p	Pyroelectric constant
p_z	Piezoelectric constant
ρ	Density
Q	Heat pulse energy
Q_0	Heat source intensity
q	Heat flux
R	Interface thermal wave reflection coefficient
Re	Real part of
R_T	Thermal contact resistance
$R(\lambda)$	Optical reflectivity at wavelength λ
σ	Thermal wavenumber
σ_{SB}	Stefan-Boltzmann constant
σ_0	DC electrical conductivity
T	Thermal wave transmission coefficient
T_0	Ambient temperature
$T(\omega)$	Periodic temperature at modulation angular frequency ω
$t_{1/2}$	Time for a transient signal to reach half maximum
τ	Mean free time, carrier life time
τ_e	Electronic time constant
θ	Infrared detector collection angle
μ	Thermal diffusion length
μ_p	Plasma diffusion length
v	Thermal wave velocity
ν	Poisson's ratio
ω	Angular modulation frequency
ω_p	Plasma frequency
x_0	Zero crossing distance (OBD)
χ	Position sensor sensitivity
Z	Thermal wave impedance

Abbreviations

FFT	Fast Fourier transform
MOR	Modulated optical reflectance
NDE	Non-destructive evaluation
OBD	Optical beam deflection
PA	Photoacoustic
PAS	Photoacoustic spectroscopy
PPE	Photo-pyroelectric
PTD	Photothermal displacement
PTR	Photothermal radiometry
PZ	Piezoelectric

1
Photothermal science

1.1 Introduction

Photothermal science encompasses a wide range of techniques and phenomena based upon the conversion of absorbed optical energy into heat. Optical energy is absorbed and eventually converted into thermal energy by an enormous number of materials – solids, liquids and gases. Although the initial absorption processes in many materials are very selective, it is common for excited electronic states in atoms or molecules to loose their excitation energy by a series of non-radiative transitions that result in a general heating of the material. Such processes are the origins of the photothermal effects and techniques that are covered in this book.

Most applications of the photothermal effect can be divided into two distinct types:

1. applications in which details of an optical absorption process are deduced by monitoring the resultant heating;
2. applications in which an optical source is used as a convenient precisely controlled form of heating.

Historically, it was the first of the two types of application that started what has become an explosion of interest in photothermal phenomena. In the mid-1970s a variant of absorption spectroscopy was developed which exploited the photothermal effect. It was named photoacoustic spectroscopy because the photothermal heating effect was detected by an indirect acoustic method. Although photoacoustic spectroscopy is discussed in considerable detail in Chapter 6, we will outline the technique here as a means of introducing the principles that are common to many of the applications of the photothermal effect.

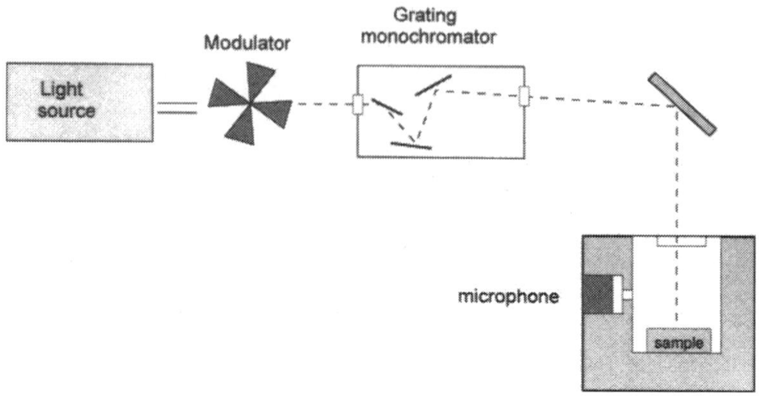

Figure 1.1 *Schematic diagram of a photoacoustic spectrometer.*

1.2 Elements of photoacoustic spectroscopy

Figure 1.1 is a schematic diagram of a photoacoustic spectrometer. The purpose of the photoacoustic cell is to monitor the temperature changes in the sample, produced by its absorption of optical energy of a wavelength determined by the grating monochromator. The cell is a small gas-tight enclosure with a sensitive acoustic microphone built into one wall. Sample temperature changes cause pressure changes in the enclosed gas that are converted to an electrical signal by the microphone. In general, the pressure changes are small. It is a well established instrumentation principle that small changes in a far larger quantity are best measured by modulating them, to allow the separate processing of the small ac component independent of the larger dc component. This principle is employed in photoacoustic spectroscopy. The light incident on the sample is modulated by a mechanical light chopper. The resulting sample heating is periodic and causes a periodic pressure change in the enclosed gas which is converted by the microphone into an ac electrical signal that may be processed using low noise narrow bandwidth techniques. The photoacoustic cell has proved to be a very sensitive, though indirect, means of detecting the sample temperature changes caused by the photothermal effect.

The modulated optical source is a common feature of all photothermal techniques and phenomena. The heating produced in all cases is non-stationary. This characteristic has led to the discovery of a wealth of phenomena associated with the interaction and propagation of modulated heat in media of all types. It has been widely recognized that such phenomena can be usefully described and understood by adopting the principles of wave physics. The term 'thermal wave' began to appear in the photothermal literature in the early part of the 1980s. A modulated light source is a very convenient means of generating thermal waves. The wavelength and propagation characteristics of a thermal wave are simply controlled by the frequency of the light source modulator. Lasers can be used to generate thermal waves at a specific point on the surface of a sample, allowing spatially resolved thermal probing. These attributes combined with the development of very sensitive thermal detection techniques led to the widespread adoption of thermal waves as a new means of 'probing' matter.

1.3 Historical perspective 1 – photoacoustic spectroscopy

Photothermal science seems to have had a remarkably late start. As mentioned above, photoacoustic spectroscopy dates from the mid-1970s yet the instrumentation it employs was available decades before. The concept of a thermal wave was treated with surprise and suspicion when it first appeared in the 1980s yet much of the basic thermal wave phenomena had been understood and put to use more than a hundred years before!

Photoacoustic spectroscopy was one of the great rediscoveries of modern science. The photoacoustic effect was first reported by Alexander Graham Bell in 1880. He discovered the effect whilst working on the 'photophone'. Bell was much ahead of his time in attempting to use a modulated beam of light to transmit a telephone call. He employed a beam of sunlight that was modulated by a voice activated mirror attached as a diaphragm to a speaking tube. The modulated light was focused onto a selenium cell incorporated in an electrical telephone circuit. The resistance of the selenium was modulated by the modulated sunlight to reproduce the transmitted speech in the telephone receiver. However, Bell discovered the selenium that emitted an audible signal when exposed to a modulated beam of light. He demonstrated that selenium and other solids attached as diaphragms to hearing tubes emitted sound when exposed to intense modulated light. He went on to show that the magnitude of the effect depended on the wavelength of the light and correctly attributed the effect to an optical absorption process, thus providing the basis for photoacoustic absorption spectroscopy. Although Bell's discovery of the photoacoustic effect gained the interest and involvement of such notable scientists as Rayleigh, Roentgen and Tyndall in the 1880s, the practical limitations of using hearing tubes consigned it to become a scientific curiosity.

The need for the use of a hearing tube was eliminated by the development of the microphone but still there was little interest in making use of the effect until the 1970s. Up to that time the only applications that can be identified as photoacoustic were in the field of gas analysis in which the absorption of light by gaseous media was detected via the resultant temperature–pressure changes using a microphone. Allan Rosencwaig deserves much of the credit for the rediscovery of the photoacoustic effect, for the development of photoacoustic spectroscopy in solids and for much of our understanding of the principles of this technique and several others that now form a part of photothermal science. The Rosencwaig–Gersho theory (Rosencwaig and Gersho 1976) of the photoacoustic effect for a sample in a photoacoustic cell provided a comprehensive theoretical framework which led to the rapid development and application of the effect. This led to the invention of numerous other detection schemes and to the current widespread interest in photothermal science.

1.4 Historical perspective 2 – thermal waves

In modern photothermal science most applications are not spectroscopic and photoacoustic cells are not used for detection. The photothermal effect is used primarily to generate a modulated heating. The temperature modulation that results is determined by the specific details of the thermal propagation within the medium concerned, in addition to its optical absorption characteristics. Because of the periodic modulation of the heat source it was natural to attempt to adopt the principles of wave physics that are so successful in explaining the propagation of other periodic phenomena, such as elastic waves. The utilization of a wave treatment of heat was in some respects another rediscovery. It can be argued that thermal waves have just as eminent a heritage as photoacoustic spectroscopy. Fourier's mathematical analysis originally appeared in *La Theorie Analytique de la Chaleur*, published in 1822, and showed that heat conduction problems in solids could be solved by expanding arbitrary imposed temperature distributions as a series of waves. He and Poisson went on to make use of the daily periodic temperature oscillations at the surface of the earth to estimate the thermal properties of the rocks near the surface. The equations they used are identical to those used to describe the thermal waves generated, at somewhat higher frequencies, in modern photothermal experiments. Ångström in 1861 devised a 'temperature-wave' method for the determination of the thermal diffusivity of a long rod. One end of the rod was heated periodically and the periodic temperature reaching the other end was measured and compared with that at the heated end. The diffusivity was shown to be given either by the attenuation of the wave or the phase lag between the periodic temperatures at the two measurement points. Again the equations employed are those of the modern thermal wave. A range of photothermal variants of Ångström's method for diffusivity measurement is described in Chapter 9.

1.5 Preview

The theoretical background for the majority of the applications of the photothermal effect is covered in Chapters 2, 3 and 5. Thermal conduction in solids, the properties of thermal waves and their basic interactions in solids are dealt with in Chapter 2. Chapter 3 is devoted to the photothermal effect; to the factors that determine the magnitude of the periodic temperature generated by optical absorption in the surface. Chapter 5 introduces pulsed, rather than periodic heating, and shows how the subsequent transient thermal response can also be understood by making use of results obtained for thermal waves.

The very wide range of experimental techniques that have been developed to exploit the photothermal effect are described and compared in Chapter 4. In addition to the photoacoustic cell, numerous ingenious methods have been devised to detect photothermally generated periodic heating. These include the use of: the piezoelectric effect; the pyroelectric effect; the emission of infrared radiation; the mirage effect; the modulation of the surface optical reflectance and the detection of the thermally induced surface distortion.

The remaining five chapters deal with five major areas of application. Photoacoustic spectroscopy is outlined in Chapter 6. The technique and the quantitative interpretation of data are explained before presenting a review of applications. The applications covered range from studies of simple powdered metals to biological studies of photosynthesis and the determination of hemoglobin content of blood. Chapter 7 introduces the field of thermal wave non-destructive evaluation. Thermal waves have excited great interest as a potentially powerful new non-destructive evaluation probe. They are being considered in competition with established techniques such as ultrasonics. The growing use of active thermography is included in this chapter. The very important applications of semiconductor evaluation and semiconductor wafer inspection are covered in Chapter 8. Particular attention is paid to semiconductor property analysis using the mirage effect and to the development of the modulated optical reflectance microscope for wafer inspection. The photothermal techniques that have been developed for thermal property measurements are discussed in Chapter 9, including the very widely used laser flash technique. Finally, the uses of photothermal methods for gas analysis and atmospheric pollution monitoring are outlined in Chapter 10.

References

Rosencwaig, A. and Gersho, A. 'Theory of the photoacoustic effect in solids', *J. Appl. Phys.*, **47** (1976) 64–9.

Further Reading

Rosencwaig, A., *Photoacoustics and photoacoustic spectroscopy*, John Wiley (1980).

2

Thermal waves

Many of the applications of the photothermal phenomena introduced in Chapter 1 depend on a clear understanding of the ways in which thermophysical properties or sub-surface structures affect heat flow. In this chapter the essential features of heat transfer are presented, followed by a detailed discussion of harmonic heat flow leading to the concept of 'thermal waves'. The wave-like properties that may be exhibited by diffusive heat flow are then further emphasized by demonstrating the occurrence of phenomena that may be regarded as: reflection, refraction, interference and scattering effects.

2.1 Basic heat transfer expressions

Whenever a temperature difference exists in a medium, or between bodies, heat transfer occurs. Heat transfer from or to a material can occur by a combination of conduction, convection and radiation processes. Figure 2.1 illustrates these three heat transfer modes schematically. In solids, thermal energy is transferred by lattice vibrations and, in the case of metals, by conduction electrons. Two mechanisms are also involved in convective heat transfer. In addition to energy transfer due to random molecular motion (diffusion) there is also the possibility of thermal energy transference by the bulk, or macroscopic, motion of a fluid in contact with the heated surface. Radiative heat transfer is a characteristic of all matter at a non-zero temperature and occurs by the emission of electromagnetic energy. The radiated energy is transported by electromagnetic waves (photons), which require no conduction medium, in contrast to the conductive and convective mechanisms of heat transfer.

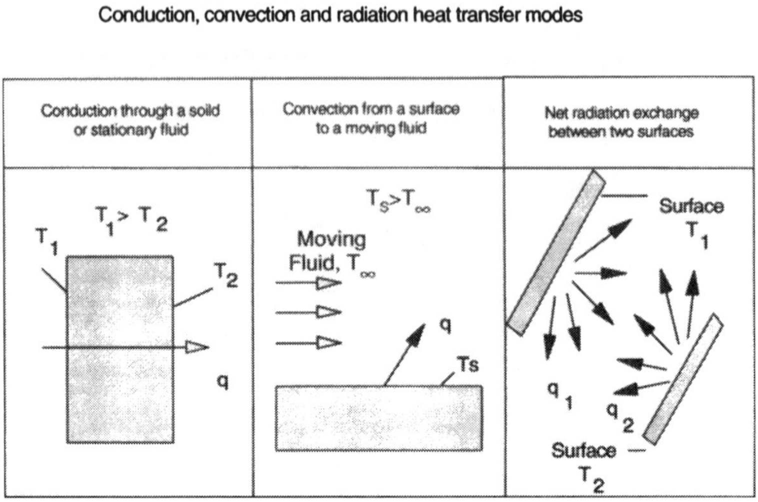

Figure 2.1 *Schematic illustration of the conduction, convection and radiation heat transfer modes.*

These heat transfer processes can be quantified by appropriate rate equations which can be used to compute the amount of energy being transferred per unit time. For heat conduction, the rate equation is known as Fourier's law. For one-dimensional heat flow, the rate equation is:

$$q_{cd} = -k\frac{\partial T}{\partial x} \tag{2.1}$$

The heat flux q_{cd} (W m^{-2}) is the rate of heat transfer per unit area in the x-direction which is proportional to the temperature gradient in this direction, $\partial T/\partial x$. The proportionality constant k is a transport property known as the thermal conductivity (W m^{-1} K^{-1}) which is a characteristic of the material. The minus sign is a consequence of the fact that heat is transferred in the direction of decreasing temperature. Note that the heat flux is a vector quantity.

For convective heat flow, the appropriate rate equation is of the form:

$$q_{cv} = h(T_s - T_\infty) \tag{2.2}$$

where q_{cv}, the convective heat flux (W m^{-2}), is proportional to the difference between the surface and fluid temperatures, T_s and T_∞ respectively. This expression is known as Newton's law of cooling, in which the proportionality constant h (W m^{-2} K^{-1}) is the convective heat transfer coefficient.

For radiative heat transfer, the maximum flux emitted by a body is given by the Stefan–Boltzmann law:

$$q_{\mathrm{rd}} = \sigma_{\mathrm{SB}} T_{\mathrm{s}}^4 \tag{2.3}$$

where T_{s} is the absolute temperature (K) of the surface and σ_{SB} is the Stefan–Boltzmann constant (5.67×10^{-8} W m^{-2} K^{-4}). Only a perfect radiator, often called a blackbody, would emit as much energy as this. The heat flux emitted by a real surface is given by:

$$q_{\mathrm{rd}} = \varepsilon \sigma_{\mathrm{SB}} T_{\mathrm{s}}^4 \tag{2.4}$$

where the emissivity ε is a parameter characterizing the radiative properties of the surface. This property indicates how efficiently the surface emits compared with an ideal radiator. The determination of the net rate at which radiation is exchanged between surfaces is generally a complex matter. However, a special case which frequently occurs in practice involves the net exchange between a small surface and a much larger surface that completely surrounds it. The net radiative heat transfer for this case is given by:

$$q = \varepsilon \sigma_{\mathrm{SB}} A (T_{\mathrm{s}}^4 - T_{\mathrm{b}}^4) \tag{2.5}$$

In this expression A is the area of the smaller surface and T_{b} is the temperature of the surroundings.

In most photothermal experiments, the induced temperature changes are small and it is only necessary to consider the effect of thermal conduction. The remainder of this chapter and those that follow will concentrate on thermal conduction effects. The other forms of heat transfer will only be included where their contributions become significant.

2.2 Differential equations of heat conduction

The analysis of most photothermal effects necessitates the determination of the temperature field in a medium resulting from conditions, such as a specific type of heating, that may be regarded as being imposed boundary conditions. The spatial and temporal temperature distribution in a medium may be determined by: the application of the principle of energy conservation to a small differential element or control volume; identifying the heat transfer

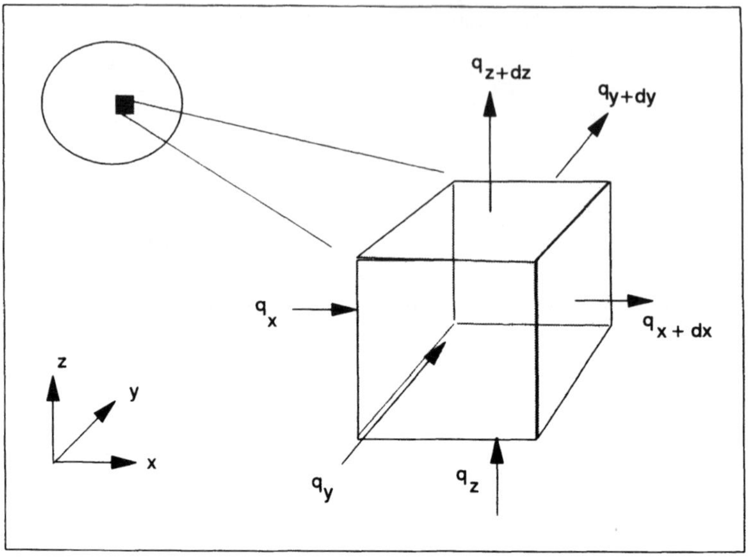

Figure 2.2 *Differential control volume element,* dx dy dz, *for conduction analysis in Cartesian co-ordinates.*

processes, and applying the appropriate rate equations. This leads to the formulation of a differential equation whose solution, for prescribed boundary conditions, provides the required temperature distribution.

With reference to Fig. 2.2, the application of the energy conservation principle to the fluxes (rate of flow with respect to time) of thermal energy into and out of the volume element, dV, of dimension dx, dy and dz leads to the expression:

$$Q_{\mathrm{in}} + Q_{\mathrm{g}} - Q_{\mathrm{out}} = Q_{\mathrm{st}} \tag{2.6}$$

where Q_{in} is the rate of energy flow into dV

$$Q_{\mathrm{in}} = q_x + q_y + q_z$$

Q_{g} is the rate of energy generation in dV

$$Q_{\mathrm{g}} = \dot{g}\,\mathrm{d}x\,\mathrm{d}y\,\mathrm{d}z$$

Q_{out} is the rate of energy flow out of dV

$$Q_{\mathrm{out}} = q_{x+\mathrm{d}x} + q_{y+\mathrm{d}y} + q_{z+\mathrm{d}z}$$

Q_{st} is the rate of energy storage in dV

$$Q_{st} = \rho c \frac{\partial T}{\partial t} \,\mathrm{d}x\,\mathrm{d}y\,\mathrm{d}z$$

ρ and c are the density and specific heat capacity respectively of the medium and $\dot{g}$ is the rate of energy generation per unit volume in the medium. Inserting these expressions in equation (2.6):

$$q_x + q_y + q_z + \dot{g}\,\mathrm{d}x\,\mathrm{d}y\,\mathrm{d}z - q_{x+\mathrm{d}x} - q_{y+\mathrm{d}y} - q_{z+\mathrm{d}z} = \rho c \frac{\partial T}{\partial t} \,\mathrm{d}x\,\mathrm{d}y\,\mathrm{d}z \tag{2.7}$$

The heat fluxes entering the cube, q_x, q_y and q_z, may be evaluated using Fourier's law,

$$q_x = -k\,\mathrm{d}y\,\mathrm{d}z\left.\frac{\partial T}{\partial x}\right|_x, \quad q_y = -k\,\mathrm{d}x\,\mathrm{d}z\left.\frac{\partial T}{\partial y}\right|_y, \quad q_z = -k\,\mathrm{d}x\,\mathrm{d}y\left.\frac{\partial T}{\partial z}\right|_z \tag{2.8}$$

and the heat fluxes exiting the cube are obtained via a Taylor expansion where, neglecting higher order terms:

$$q_{x+\mathrm{d}x} = q_x + \frac{\partial q_x}{\partial x}dx, \quad q_{y+\mathrm{d}y} = q_y + \frac{\partial q_y}{\partial y}\mathrm{d}y, \quad q_{z+\mathrm{d}z} = q_z + \frac{\partial q_z}{\partial z}\mathrm{d}z \tag{2.9}$$

Substituting these equations into the energy conservation equation, (2.7), we then obtain:

$$-\frac{\partial q_x}{\partial x}\mathrm{d}x - \frac{\partial q_y}{\partial y}\mathrm{d}y - \frac{\partial q_z}{\partial z}\mathrm{d}z + g\,\mathrm{d}x\,\mathrm{d}y\,\mathrm{d}z = \rho c\frac{\partial T}{\partial t}\mathrm{d}x\,\mathrm{d}y\,\mathrm{d}z \tag{2.10}$$

which becomes:

$$\frac{\partial}{\partial x}\left(k\,\mathrm{d}y\,\mathrm{d}z\frac{\partial T}{\partial x}\right)\mathrm{d}x + \frac{\partial}{\partial y}\left(k\,\mathrm{d}x\,\mathrm{d}z\frac{\partial T}{\partial y}\right)\mathrm{d}y + \frac{\partial}{\partial z}\left(k\,\mathrm{d}x\,\mathrm{d}y\frac{\partial T}{\partial z}\right)\mathrm{d}z$$
$$+\dot{g}\,\mathrm{d}x\,\mathrm{d}y\,\mathrm{d}z = \rho c\frac{\partial T}{\partial t}\mathrm{d}x\,\mathrm{d}y\,\mathrm{d}z \tag{2.11}$$

This is the general form, in Cartesian co-ordinates, of the heat diffusion equation which provides the basic tool for heat conduction analysis. From its solution, we can obtain the temperature distribution *T(x,y,z)* as a function of time. The heat diffusion equation is simply an expression of the energy conservation principle.

The heat diffusion equation states that:

> at any point in a medium the net rate of heat conduction into a volume element plus the heat generation rate within the element must equal the rate of change of thermal energy stored at that point in the medium.

The heat differential equation can be further simplified to produce four special cases:

1. If it can be assumed that the thermal conductivity of the material is isotropic,

$$\frac{\partial^2 T}{\partial x^2}+\frac{\partial^2 T}{\partial y^2}+\frac{\partial^2 T}{\partial z^2}+\frac{\dot{g}}{k}=\frac{1}{\alpha}\frac{\partial T}{\partial t} \tag{2.12}$$

where $\alpha = k/\rho c$ is the thermal diffusivity ($m^2\ s^{-1}$). This important thermophysical property is the ratio of the thermal conductivity and the thermal capacitance, ρc. The thermal diffusivity indicates the rate at which heat is distributed in a material and this rate depends not only on the thermal conductivity but also on the rate at which energy can be stored.

2. No internal heat generation:

$$\frac{\partial^2 T}{\partial x^2}+\frac{\partial^2 T}{\partial y^2}+\frac{\partial^2 T}{\partial z^2}=\frac{1}{\alpha}\frac{\partial T}{\partial t} \tag{2.13}$$

this expression if often referred to as the Fourier equation or the diffusion equation.

3. Steady-state with internal heat generation:

$$\frac{\partial^2 T}{\partial x^2}+\frac{\partial^2 T}{\partial y^2}+\frac{\partial^2 T}{\partial z^2}+\frac{g(x,y,z)}{k}=0 \tag{2.14}$$

which is the familiar Poisson equation.

4. Steady-state and no internal heat generation:

$$\frac{\partial^2 T}{\partial x^2}+\frac{\partial^2 T}{\partial y^2}+\frac{\partial^2 T}{\partial z^2}=0 \tag{2.15}$$

which is the Laplace equation.

2.3 Thermal wave generation and propagation

Consider an isotropic homogeneous semi-infinite medium whose surface is subjected to plane harmonic heating, of the form $(Q_0/2)[1 + \cos(\omega t)]$, where Q_0 is the source intensity, ω is the angular modulation frequency of the heat source and t is time. If the heated surface is taken to occupy the y–z plane at $x = 0$, the temperature distribution within the solid can be obtained by solving a one-dimensional heat diffusion equation in x and t:

$$\frac{\partial^2 T}{\partial x^2} - \frac{1}{\alpha}\frac{\partial T}{\partial t} = 0 \qquad x < 0, t > 0 \tag{2.16}$$

subject to the boundary condition that the periodic thermal energy applied to the surface is dissipated by conduction into the solid:

$$-k\frac{\partial T}{\partial x} = \frac{Q_0}{2}[1 + \cos(\omega t)] = \mathrm{Re}\left\{\frac{Q_0}{2}[1 + \exp(\mathrm{j}\omega t)]\right\} \quad x=0,\ t>0 \tag{2.17}$$

in which Re stands for 'the real part of' and $\surd(-1)$.

The heating divides into two parts, $Q_0/2$ and $(Q_0/2)\exp(\mathrm{j}\omega t)$, which produce a dc temperature increase and an ac thermal modulation respectively. We are primarily interested in the ac component and will omit the dc component in the following solution. To solve equation (2.16), let us assume the periodic component has a solution of the form:

$$T(x,t) = \mathrm{Re}(T(x)\exp(\mathrm{j}\omega t)) \tag{2.18}$$

Omitting the 'Re' symbol, substituting equation (2.18) in equation (2.16), we obtain

$$\mathrm{e}^{\mathrm{j}\omega t}\left(\frac{\mathrm{d}^2 T(x)}{\mathrm{d}x^2} - \frac{\mathrm{j}\omega}{\alpha}T(x)\right) = 0 \tag{2.19}$$

Discarding the exponential time factor, the general solution for the spatial dependence of the temperature may be written in the form:

$$T(x) = A\exp(-\sigma x) + B\exp(\sigma x) \qquad \sigma = (1+\mathrm{j})\left(\frac{\omega}{2\alpha}\right)^{1/2} \tag{2.20}$$

where A and B are arbitrary constants. To evaluate these constants we note that as x tends to infinity $T(x)$ must be finite and therefore the constant B is zero. The expression for A is evaluated by applying the flux continuity boundary condition at the sample surface, x=0:

$$\frac{Q_0}{2} = -k\frac{\partial T(x)}{\partial x} = (-k)(-\sigma)A\exp(-\sigma x) \qquad (2.21)$$

From which we obtain $A = Q_0/2k\sigma$, and thus the full solution:

$$\begin{aligned} T(x,t) &= \frac{Q_0}{2k\sigma}\ \exp(-\sigma x + \mathrm{j}\omega t) \\ &= \frac{Q_0}{2\sqrt{\rho c k \omega}}\ \exp\left(-x\sqrt{\frac{\omega}{2\alpha}}\right)\exp\mathrm{j}\left(\omega t - x\sqrt{\frac{\omega}{2\alpha}} - \frac{\pi}{4}\right) \end{aligned} \qquad (2.22)$$

This is a similar expression to that obtained for the amplitude of an electromagnetic wave in the surface skin depth of an electrical conductor, such as a metal. In Fig. 2.3 the amplitude and phase of the 'thermal wave' are shown plotted as a function of depth below the surface. The exponential but oscillatory decay of the temperature amplitude and the linear phase variation with sample depth should be noted. Several wave-like features of this temperature solution can be identified.

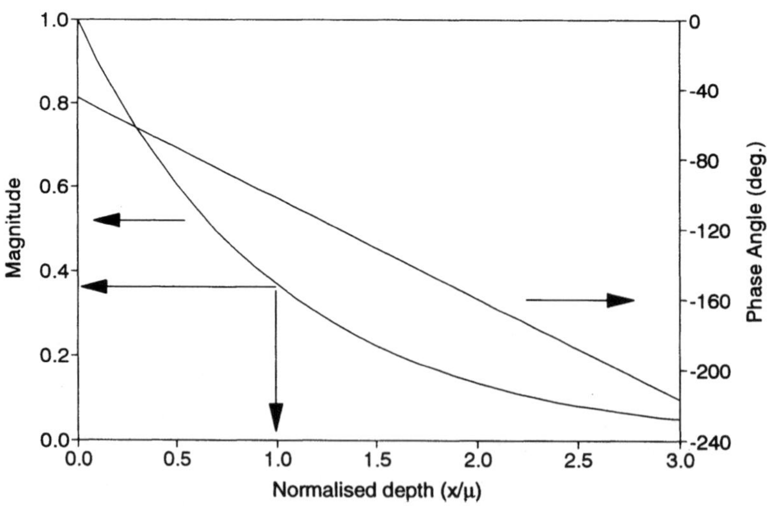

Figure 2.3 *Thermal wave amplitude and phase shown as a function of reduced depth x/μ.*

1. Like normal propagating waves, the thermal wave has an oscillatory spatial dependence of the form exp(j*kx*), with a wave vector given by:

$$\mathrm{Re}(\sigma) = \frac{1}{\mu} = \sqrt{\frac{\omega}{2\alpha}}$$

in which the parameter μ, called the thermal diffusion length, a quantity analogous to the electromagnetic skin depth, is defined as $\sqrt{(2\alpha/\omega)}$. Equation (2.22) then becomes:

$$T(x,t) = \frac{Q_0}{2\sqrt{\rho c k \omega}} \exp(-x/\mu) \exp \mathrm{j}\left[\omega t - x/\mu - \frac{\pi}{4}\right] \quad (2.23)$$

2. Unlike normal propagating waves, a thermal wave is very heavily damped with a decay constant equal to the thermal diffusion length. Table 2.1 lists the thermal properties and thermal diffusion lengths for various common materials. From the definition of the thermal diffusion length it can be seen that thermal waves will propagate deep into a solid if the material has a high thermal diffusivity or if the thermal wave frequency is low. The thermal diffusion lengths of aluminum, stainless steel, air and Bakelite in the frequency range

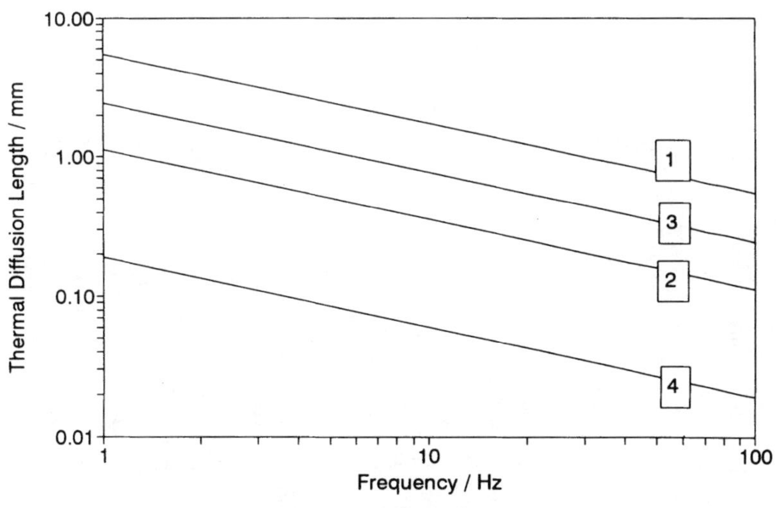

Figure 2.4 *A plot of the thermal diffusion length versus frequency for: (1) aluminum, (2) stainless steel, (3) air and (4) Bakelite.*

Table 2.1 Thermophysical properties of some common materials

Material	Density ($kg\ m^{-3}$)	Specific heat ($J\ kg^{-1}K^{-1}$)	Thermal conductivity ($W\ m^{-1}\ K^{-1}$)	Thermal diffusivity ($\times 10^{-6}\ m^2\ s^{-1}$)	Thermal effusivity ($Ws^{1/2}\ m^{-2}\ K^{-1}$)	Thermal diffusion length mm 1 Hz	mm 25 Hz	mm 100 Hz
Metals								
Pure aluminum	2700	945	238	93.28	24 642.59	5.4490	1.0898	0.5449
Al alloy 2024-T6	2770	875	177	73.03	20 712.41	4.8213	0.9643	0.4821
Chromium	7160	449	93.7	29.15	17 355.99	3.0459	0.6092	0.3046
Pure copper	8933	385	401	116.60	37 136.52	6.0921	1.2184	0.6092
Bronze 90% Cu, 10% Al	8800	420	52	14.07	13 863.33	2.1162	0.4232	0.2116
Gold	19 300	129	317	127.32	28 093.32	6.3662	1.2732	0.6366
Pure iron	7870	447	80.2	22.80	16 796.87	2.6938	0.5388	0.2694
Steel AISI 1010	7832	434	63.9	18.80	14 737.77	2.4462	0.4892	0.2446
Steel AISI 316	8238	468	13.4	3.48	7187.64	1.0518	0.2104	0.1052
Lead	11 340	129	35.3	24.13	7186.03	2.7715	0.5543	0.2771
Magnesium	1740	1024	156	87.55	16 671.97	5.2791	1.0558	0.5279
Molybdenum	10 240	251	138	53.70	18 833.30	4.1344	0.8269	0.4134
Pure nickel	8900	444	90.7	22.95	18 931.72	2.7030	0.5406	0.2703
Inconel X-750	8510	439	11.7	3.13	6611.35	0.9984	0.1997	0.0998
Silver	10 500	235	429	173.86	32 535.48	7.4392	1.4878	0.7439
Titanium	4500	522	21.9	9.32	7172.38	1.7227	0.3445	0.1723
Tungsten	19 300	132	174	68.30	21 054.27	4.6627	0.9325	0.4663

Table 2.1 *(cont.)*

Material	Density ($kg\ m^{-3}$)	Specific heat ($J\ kg^{-1}K^{-1}$)	Thermal conductivity ($W\ m^{-1}\ K^{-1}$)	Thermal diffusivity ($\times 10^{-6}\ m^2\ s^{-1}$)	Thermal effusivity ($Ws^{1/2}\ m^{-2}\ K^{-1}$)	Thermal diffusion length mm 1 Hz	mm 25 Hz	mm 100 Hz
Zinc	7140	389	116	41.76	17 949.52	3.6461	0.7292	0.3646
Zirconium	6570	278	22.7	12.43	6438.99	1.9890	0.3978	0.1989
Non-metals								
Aluminum oxide	3970	765	46	15.15	11 819.66	2.1957	0.4391	0.2196
Beryllium oxide	3000	1030	272	88.03	28 991.03	5.2933	1.0587	0.5293
Diamond IIa	3500	509	2300	1291.05	64 011.33	20.2720	4.0544	2.0272
Silicon carbide	3160	675	490	229.72	32 329.09	8.5512	1.7102	0.8551
Silicon nitride	2400	691	16	9.65	5151.16	1.7524	0.3505	0.1752
Germanium	5360	322	59.9	34.71	10 167.72	3.3237	0.6647	0.3324
Silicon	2330	712	148	89.21	15 669.27	5.3289	1.0658	0.5329
Quartz glass	2210	730	1.4	0.87	1502.87	0.5256	0.1051	0.0526
Bakelite	1270	1590	0.23	0.11	681.50	0.1904	0.0381	0.0190
Rubber (soft)	1100	2010	0.13	0.06	536.12	0.1368	0.0274	0.0137
Wood (oak)	545	2385	0.17	0.13	470.07	0.2040	0.0408	0.0204
Lubricating oil	845	2020	0.14	0.08	488.84	0.1616	0.0323	0.0162
Air	1.16	1007	0.026	22.26	5.51	2.6618	0.5324	0.2662

1–100 Hz are shown in Fig. 2.4. The thermal diffusivity span for these materials is from 10^{-5} to 10^{-7} m^2 s^{-1}. The thermal diffusion length is a key parameter because it provides an indication of the depth to which a thermal wave technique will be effective.

3. Thermal waves are highly dispersive, their phase velocity is defined in the usual way as:

$$v = \omega\mu = \sqrt{2\alpha\omega} \tag{2.24}$$

indicating that the high frequency thermal waves propagate faster than low frequency thermal waves. This has important implications for transient thermal wave materials inspection (Chapter 5) where an infinite number of different frequency components may be present.

4. There is a progressive phase variation between the temperature at the surface $x = 0$ and at a point x on the propagating thermal wave:

$$\varphi = -\frac{x}{\mu} - \frac{\pi}{4} \tag{2.25}$$

At the surface the phase lag is –45° between the heat source and the resulting surface temperature.

5. The thermal wave impedance is defined as the ratio of the temperature to the heat flux density, making an analogy with electrical impedance in which current density is analogous to heat flux density and electric field to temperature:

$$Z = \frac{1}{k\sigma} = \frac{1}{\sqrt{\mathrm{j}\omega\rho ck}} \tag{2.26}$$

The thermal wave impedance determines the magnitude of the thermal wave at the sample surface. The quantity $\sqrt{(\rho ck)}$ in the impedance equation is known as the thermal effusivity or inertia. Low values of thermal effusivity lead to high amplitudes of the surface temperature. Thermal effusivities are shown alongside thermal diffusivities in Table 2.1. In general, a high diffusivity material also has a high effusivity but there are exceptions to this rule. The most important exception is air which has a high diffusivity because its very low conductivity is balanced by its equally low density. It has a very low effusivity, however, because effusivity is determined by the product of density and conductivity. This very low effusivity is an essential characteristic exploited in the many non-destructive evaluation applications of thermal waves which will be discussed at length in Chapter 7.

2.4 Reflection and refraction of thermal waves

Thermal waves are reflected and refracted at interfaces in the same way as electromagnetic or acoustic waves. The thermal wave equation can be used to derive the thermal wave reflection and transmission coefficients at a thermal boundary. We will assume that the boundary separating two media (1,2) is the plane $x = 0$, and that the incident, reflected and transmitted plane thermal waves make angles θ_i, θ_r and θ_t with respect to the x-axis, as shown in Fig. 2.5. Let the expressions for the incident, reflected and transmitted plane thermal waves, T_i, T_r and T_t be given by:

$$\begin{aligned} T_i &= A\exp(-\sigma_1 x\cos\theta_i - \sigma_1 y\sin\theta_i + j\omega t) \\ T_r &= AR\exp(\sigma_1 x\cos\theta_r - \sigma_1 y\sin\theta_r + j\omega t) \\ T_t &= AT\exp(-\sigma_2 x\cos\theta_i - \sigma_2 y\sin\theta_t + j\omega t) \end{aligned} \tag{2.27}$$

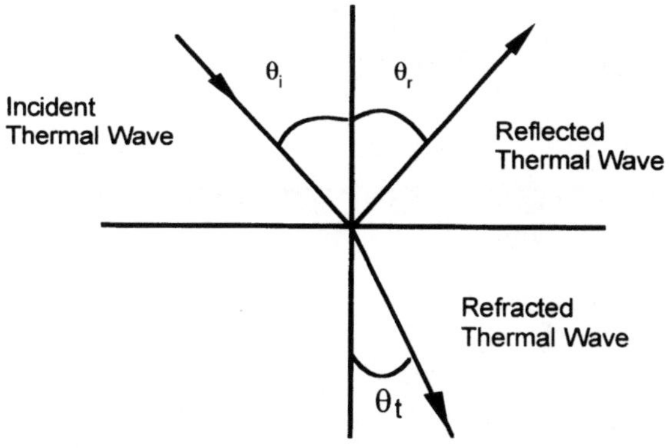

Figure 2.5 *Reflection and refraction of thermal waves at a media boundary. Coating/substrate sample system.*

where A is the initial amplitude of the incident thermal wave; R and T are the thermal wave reflection and transmission coefficients at the interface x=0. For the continuity of temperature at the interface x=0, we have:

$$A\exp(-\sigma_1 y\sin\theta_i) + AR\exp(-\sigma_1 y\sin\theta_r) = AT\exp(-\sigma_2 y\sin\theta_t) \quad (2.28)$$

As this condition must be true for all y positions across the interface plane, the exponents in the above equation must be all equal. Thus we have

$$\sigma_1 y\sin(\theta_i) = \sigma_1 y\sin(\theta_r) = \sigma_2 y\sin(\theta_t) \quad (2.29)$$

from which the laws of reflection and refraction follow:

$$\theta_i = \theta_r \quad \text{law of reflection} \quad (2.30)$$

$$\sigma_i \sin(\theta_i) = \sigma_2 \sin(\theta_t) \quad \text{law of refraction} \quad (2.31)$$

For continuity of flux at the interface the following equation is obtained by differentiating each component of equation 2.27:

$$k_1\sigma_1\cos(\theta_i) - Rk_1\sigma_1\cos(\theta_r) = Tk_2\sigma_2\cos(\theta_t) \quad (2.32)$$

as θ_i = θ_r and $T = (1 + R)$ (at $y = 0$), from equation (2.28), the thermal wave reflection and transmission coefficients can be written as:

$$R = \frac{k_1\sigma_1\cos(\theta_i) - k_2\sigma_2\cos(\theta_t)}{k_1\sigma_1\cos(\theta_i) + k_2\sigma_2\cos(\theta_t)} = \frac{\cos(\theta_i) - b\cos(\theta_t)}{\cos(\theta_i) + b\cos(\theta_t)} \quad (2.33)$$

$$T = \frac{2k_1\sigma_1\cos(\theta_i)}{k_1\sigma_1\cos(\theta_i) + k_2\sigma_2\cos(\theta_t)} = \frac{2\cos(\theta_i)}{\cos(\theta_i) + b\cos(\theta_t)} \quad (2.34)$$

where

$$b = k_2\sigma_2 / k_1\sigma_1 = \sqrt{k_2\rho_2 c_2} / \sqrt{k_1\rho_1 c_1}$$

For normal incidence, θ_i = θ_t = 0, the reflection and transmission coefficients at x = 0 reduce to:

$$R = \frac{1-b}{1+b}, \quad T = \frac{2}{1+b} \quad (2.35)$$

Thus the reflection and transmission coefficient expressions for thermal waves have exactly the same form as the equivalent expressions for electromagnetic or acoustic waves. The magnitudes of these coefficients are determined by the parameter b which is seen to amount to the ratio of the effusivities of the two media and which may be regarded as being a measure of the thermal mismatch between the two media.

2.5 Thermal wave interference in a thin layer

Having established the basic properties of thermal waves and their interactions at interfaces it becomes possible to adopt the well known components of optical theory to analyze thermal wave effects. Thermal waves have proved to be particularly suitable for investigating the properties of coatings and when considering these it is natural to introduce the concept of interference. Consider the propagation of plane thermal waves in a thin layer of material of thickness L. We assume that the thermal properties of all the media

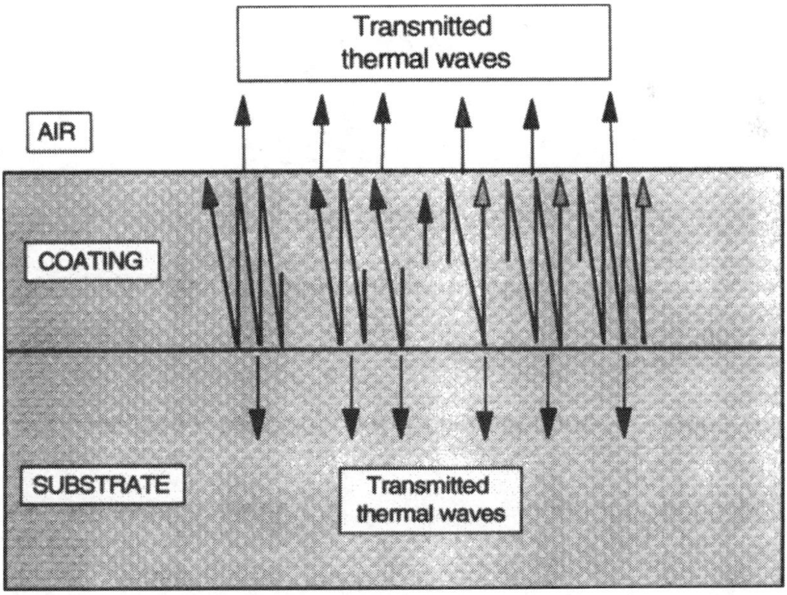

Figure 2.6 *Thermal wave components in a coating layer illustrating the origination of thermal wave interference in the layer.*

concerned are homogeneous. The subscripts c, b, and g will be used here to refer to the coating, substrate and gas media, respectively. The substrate layer is assumed to be thermally thick, i.e. much thicker than the thermal diffusion length. The coating surface is exposed to air. Thermal waves are typically generated in a coating layer by the absorption of plane periodic optical energy. In general, thermal waves of amplitude A will be generated at differing depths x beneath the surface. Details of photothermal generation, including the dependence of A on optical and thermal properties, are given in Chapter 3.

The waves generated near the surface will propagate towards the coating–substrate interface and back towards the coating surface as illustrated in Fig. 2.6. On striking the boundaries the waves will be partially reflected and transmitted. Thermal wave interference will then occur between the reflected and incident wave trains. The surface temperature is obtained by summing all the waves arriving at the coating surface. For waves reflected first from the coating–air boundary the contribution is:

$$\begin{aligned} &AT_g\left[1 + R_g R_b \exp(-2\sigma_c L_c) + \ldots + (R_g R_b)^n \exp(-2n\sigma_c L_c)\right] \\ &= AT_g \sum_{n=0}^{n=\infty} (R_g R_b)^n \exp(-2n\sigma_c L_c) \\ &= \frac{AT_g}{1 - R_b R_g \exp(-2\sigma_c L_c)} \end{aligned} \tag{2.36}$$

and for waves reflected first from the coating–substrate boundary the contribution is:

$$\begin{aligned} &[AT_g R_b \exp(-2\sigma_c L_c) + R_b (R_g R_b) \exp(-4\sigma_c L_c) + \ldots \\ &+ R_b (R_g R_b)^n \exp(-2n\sigma_c L_c)] \\ &= AT_g R_b \exp(-2\sigma_c L_c) \sum_{n=0}^{n=\infty} (R_g R_b)^n \exp(-2n\sigma_c L_c) \\ &= \frac{AT_g R_b \exp(-2\sigma_c L_c)}{1 - R_b R_g \exp(-2\sigma_c L_c)} \end{aligned} \tag{2.37}$$

R_b and R_g are the thermal wave reflection coefficients at the coating–substrate interface and the coating–air interface respectively. T_g is the thermal wave transmission coefficient at the coating–air interface. The values of these coefficients can be obtained from the definitions of the reflection and transmission coefficients, equations (2.35). The reflection and transmission coefficients for the coating–air interface, R_g and T_g, have values close to 1 and 2, respectively, because of the very small magnitude of b, in equations (2.35), resulting from the very low effusivity of air.

The sum of the two terms arising from the surface heating gives the required thermal wave interference expression for the sample surface temperature:

$$T(x=0) = AT_g\left[\frac{1+R_b\exp(-2\sigma_c L_c)}{1-R_b R_g\exp(-2\sigma_c L_c)}\right] \tag{2.38}$$

This interference expression for surface generated thermal waves in a thin layer was first obtained in this way by Bennett and Patty (1982) and it forms a basis for the quantitative analysis of thermal wave effects in absorbing or opaque thin coatings on semi-infinite non-absorbing substrates. Its practical value is that it provides a means of obtaining the thermal properties or the layer thickness values from experimental data.

In this expression the temperature at the surface is complex because the thermal wave number σ_c is a complex quantity. This complex surface temperature may be interpreted physically as arising from the fact that the temperature achieved at the surface by the applied heating is generally out of phase with the heat source. The complex expression may be manipulated in the usual fashion to provide a polar co-ordinate representation of the amplitude and the phase of the surface temperature with respect to the applied periodic heating:

amplitude =

$$\frac{\sqrt{\left[1+R_b(1-R_g)\,e^{-x}\cos x - R_b^2 R_g\,e^{-2x}\right]^2+\left[R_b(1+R_g)\,e^{-x}\sin x\right]^2}}{\left[1-R_bR_g\,e^{-x}\cos x\right]^2+\left[R_bR_g\,e^{-x}\sin x\right]^2} \tag{2.39}$$

$$\text{phase} = \tan^{-1}\left[\frac{-(1+R_g)R_b\,e^{-x}\sin x}{1+(1-R_g)R_b\,e^{-x}\cos x - R_b^2R_g\,e^{-2x}}\right] \tag{2.40}$$

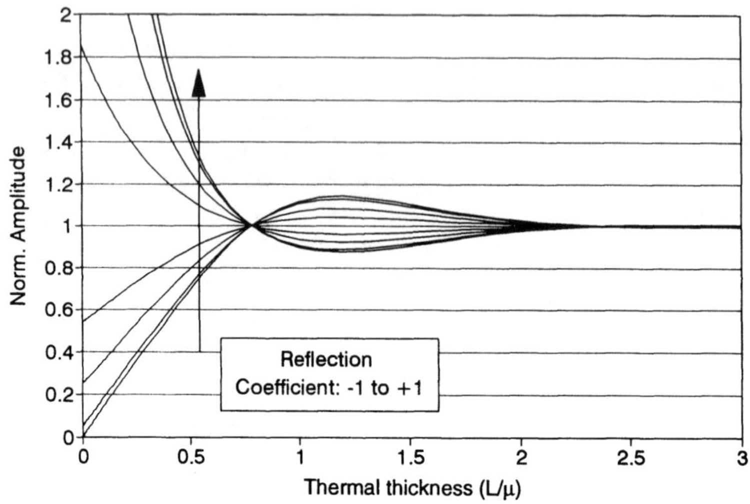

Figure 2.7 *Normalized signal amplitude variation with normalized coating thickness for various coating/substrate reflection coefficients (±1, 0.9, 0.6 and 0.3).*

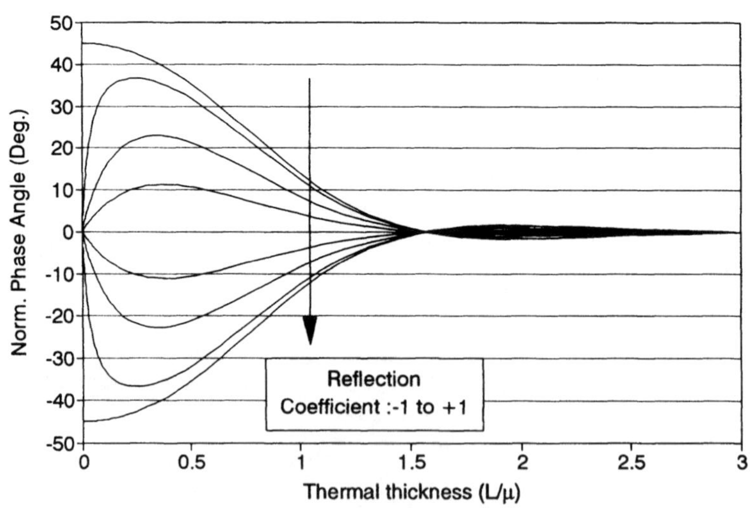

Figure 2.8 *Normalized phase angle variation with normalized coating thickness for various coating/substrate reflection coefficients (±1, 0.9, 0.6 and 0.3).*

where $x = 2L_c/\mu_c$. The functional dependencies of equations (2.39) and (2.40) are shown in Figs 2.7 and 2.8. In these figures the thermal wave interference effects are shown plotted against the reduced coating thickness L_c/μ_c. A number of different curves are shown in each figure to demonstrate the dependence of the effects on the coating–substrate reflection coefficient. The bulk of the interference effects occur where coating thickness is less than a thermal diffusion length because of the heavily damped nature of thermal waves. Consequently, the determination of the modulation frequency that satisfies this condition is an essential feature for any application exploiting thermal wave interference. Further discussion of the utilization of these thermal wave interference characteristics for coating thickness assessment or thermal property determination will be left to Chapters 7 and 9.

2.6 Reflecting power of an air gap and an imperfect interface

The reflection of a thermal wave from a thin buried air layer such as a crack is of particular practical interest and it differs substantially from the reflection process dealt with in Section 2.4. In Section 2.4 the thermal wave reflection coefficient was obtained for a boundary formed by two media in contact. This coefficient was independent of the frequency of the thermal wave. Its value was determined by the thermal mismatch between the two media, quantified by the parameter *b*.

When the interface is formed by another layer, for example an air gap, thermal wave interference occurs in this layer as well. The effective reflection coefficient of the layer can be shown to be:

$$\Gamma = \frac{\Gamma_1 + \Gamma_2 \exp(-2\sigma_2 L_2)}{1 + \Gamma_1 \Gamma_2 \exp(-2\sigma_2 L_2)} \tag{2.41}$$

where

$$\Gamma_1 = \frac{1 - b_{21}}{1 + b_{21}}, \quad \Gamma_2 = \frac{1 - b_{32}}{1 + b_{32}}, \quad b_{i+1/i} = \frac{k_{i+1}\sigma_{i+1}}{k_i \sigma_i} \tag{2.42}$$

The subscripts 1, 2 and 3 refer to the media 1, 2 and 3 respectively, as illustrated in Fig. 2.9, and Γ_1 and Γ_2 are the thermal

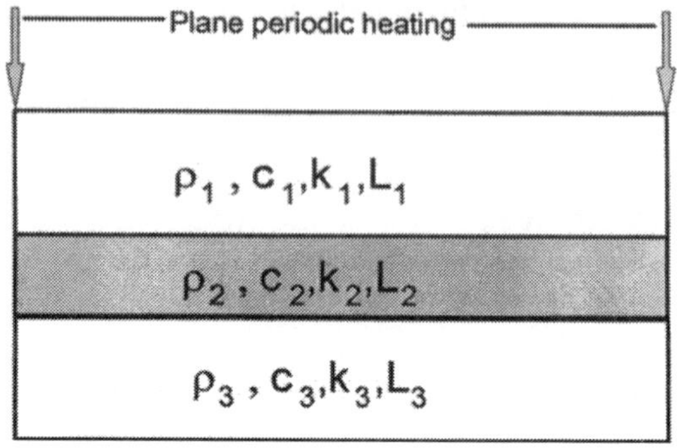

Figure 2.9 *Coating–air gap–substrate sample system.*

wave reflection coefficients for the 1/2 and 2/3 media boundaries, respectively. L_2 is the thickness of the second layer.

The thermal wave reflection coefficient of the second layer is complex and frequency dependent. The magnitude of this coefficient has important implications for the use of thermal waves to detect sub-surface features such as air gaps and cracks. In Fig. 2.10, the predicted reflectivity of an air gap in stainless steel is shown plotted as a function of air-gap thickness for modulation frequencies

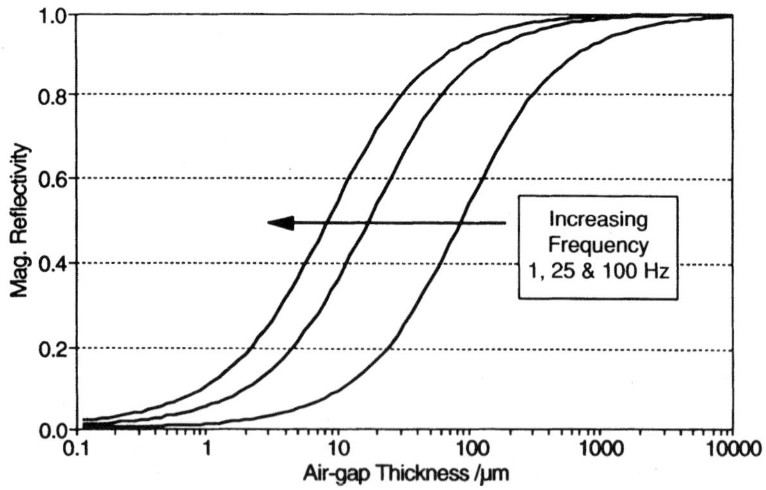

Figure 2.10 *Thermal wave reflectivity of an air gap in stainless steel at 1, 25 and 100 Hz.*

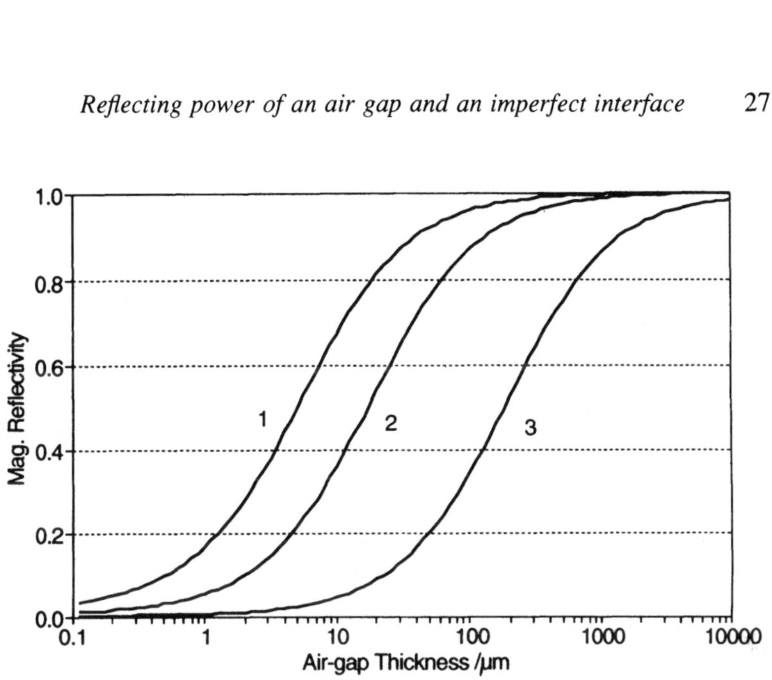

Figure 2.11 *Thermal wave reflectivity of an air gap in: (1) aluminum, (2) stainless steel and (3) Bakelite at 25 Hz.*

of 1, 25 and 100 Hz. As the thickness of the air gap increases the reflectivity increases and approaches unity (fully reflecting). Increasing the modulation frequency reduces the thermal wave length and increases the reflecting power of a particular layer of air.

If the air gap is embedded in a different material, this may affect the magnitude of the reflection coefficient. Figure 2.11 shows the reflectivities expected for air gaps in high and low effusivity materials. In a high effusivity material such as aluminum the air layer presents a higher thermal contrast and is much more reflective than in a low effusivity material such as Bakelite, suggesting that the detectability of an air-gap defect will be good in high thermal effusivity materials but also placing limitations on the prospects for thermal wave techniques for other applications.

The above model of an air gap defect is limited to situations in which distinct layers are present. In other applications where components are welded or coated or where rough surfaces are in contact this model which assumes temperature continuity at the interface is invalid. In these cases it is more appropriate to assume a temperature 'jump' at the interface, and to employ a thermal resistance model. The model used assumes flux continuity across the interface and a thermal discontinuity:

$$-k_1 \frac{\partial T_1}{\partial x} = -k_2 \frac{\partial T_2}{\partial x}, \quad T_2 - T_1 = R_T \frac{\partial T_2}{\mathrm{d}x} \tag{2.43}$$

where R_T is the interface thermal contact resistance. The thermal wave reflection coefficient for this type of an interface is then given by:

$$\Gamma = \frac{1 - b_{21} + R_T k_2 \sigma_2}{1 + b_{21} + R_T k_2 \sigma_2} \tag{2.44}$$

In Figs 2.12 and 2.13 plots equivalent to Figs 2.10 and 2.11 are presented showing the variation of the thermal wave reflection coefficient with thermal contact resistance coefficient and thermal wave frequency. It is clear that interfaces of this kind have very similar properties to air gaps.

2.7 Thermal wave scattering

The thermal wave interactions that have been dealt with so far are at semi-infinite boundaries between media. In reality, of course, these boundaries are finite in extent but provided their extent is

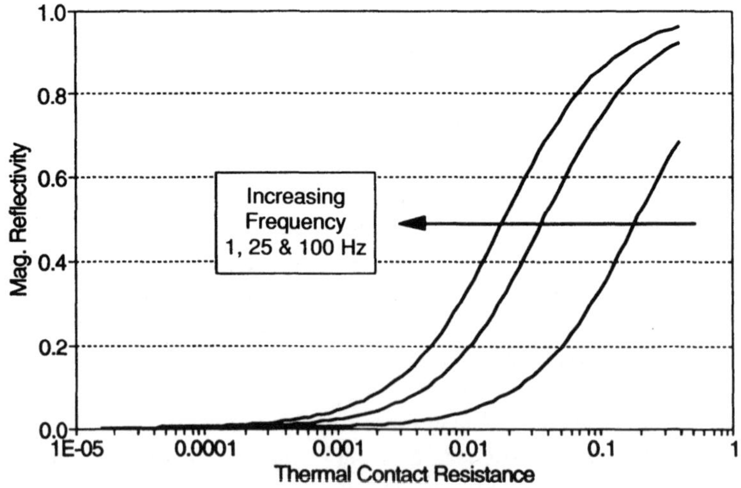

Figure 2.12 *Thermal wave reflectivity of contact resistance interface in stainless steel at 1, 25 and 100 Hz.*

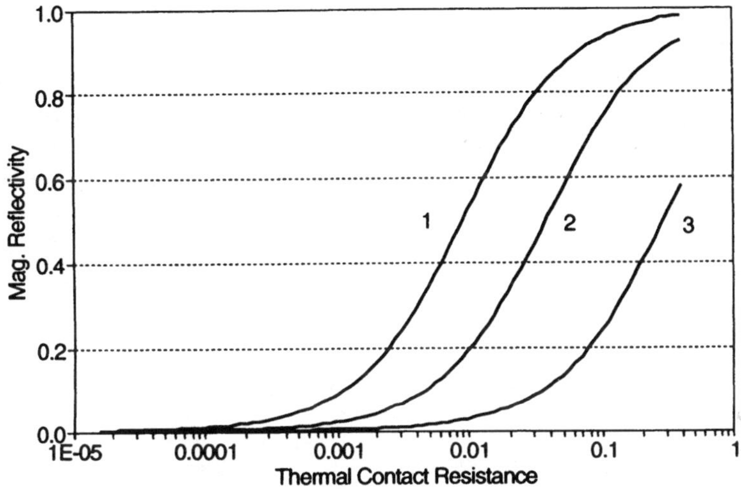

Figure 2.13 *Thermal wave reflectivity of contact resistance interface in: (1) aluminum, (2) stainless steel and (3) Bakelite at 25 Hz.*

large compared with the thermal wave length, the semi-infinite treatment can be regarded as being a good approximation. At the edges of these boundaries and at smaller features the semi-infinite treatment is inadequate. At edges and at small thermal discontinuities the thermal wave interactions can be analyzed using techniques borrowed from optical diffraction or scattering theory. To illustrate this thermal wave scattering we analyze the interaction of thermal waves with a finite size subsurface 'defect' in a homogeneous medium. The defect is taken to be a disk of radius a at depth Z below the sample surface. The surface of the sample is heated by a plane periodic heat source of angular frequency ω. Figure 2.14 shows a schematic illustration of the sample. It is of interest to examine the ways in which the surface temperature above the defect depends on the defect size and depth. Thermal waves generated at the sample surface propagate into the solid until their flow is perturbed by the disk defect. Physically the situation is similar to the reflection of light by a 'small' disk. Where the dimensions of the disk are comparable to the wavelength, a Huygens' construction may be employed to compute the reflected image. Here the defect surface is taken to comprise a set of secondary wave sources from each of which thermal waves are generated and propagate in all directions above the defect plane. A Huygens' calculation can

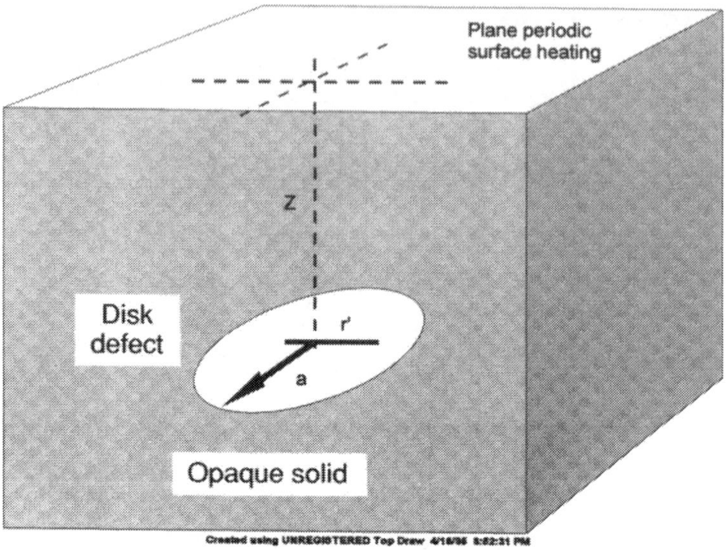

Figure 2.14 *Geometry employed in the analysis of thermal wave scattering from a subsurface disk defect.*

then be employed to sum all these scattered thermal waves at the sample surface. Using the notation shown in Fig. 2.14, the surface temperature *T(r)* per unit heat flux is given by:

$$T(r) = \frac{1}{k\sigma}\left[1 - \int T(r')\, k\left(\frac{\partial}{\partial n} G(r,r')\right) dA\right] \qquad (2.45)$$

where $T(r')$ is the temperature at the point r' on the surface of the defect and $G(r, r')$ is the Green's function which determines the thermal disturbance at r due to a point source at r', and $\partial/\partial n$ is the normal derivative at the surface. Over the center of the defect at $r = 0$ this reduces to: $T(0) =$

$$= \frac{1}{k\sigma}\left\{1 - \int_0^{2\pi} d\theta \int_0^a r\, dr \exp(-\sigma Z) k\left[\frac{\partial}{\partial Z}\left(\frac{2\exp(-\sigma\sqrt{a^2+Z^2})}{4\pi k\sqrt{a^2+Z^2}}\right)\right]\right\}$$

$$= \frac{1}{k\sigma}\left\{1 - \frac{2\exp(-\sigma Z)}{k\sigma}\left[\exp(-\sigma Z) - \frac{Z}{\sqrt{a^2+Z^2}} \exp(-\sigma\sqrt{a^2+Z^2})\right]\right\}$$

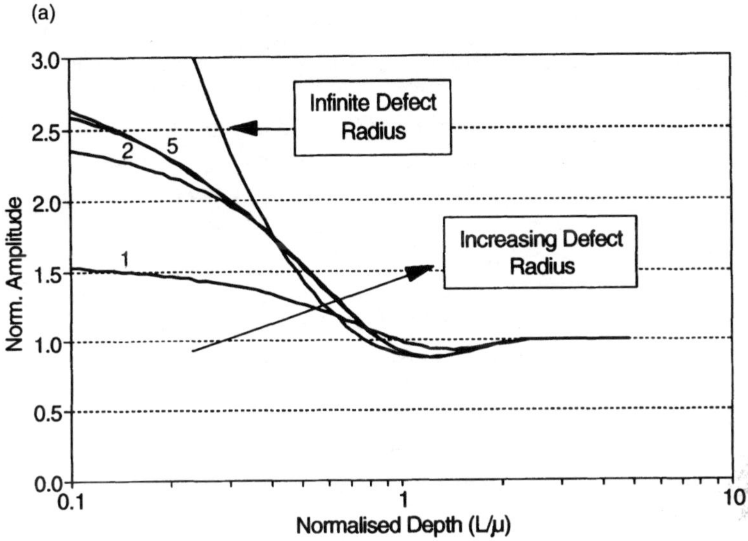

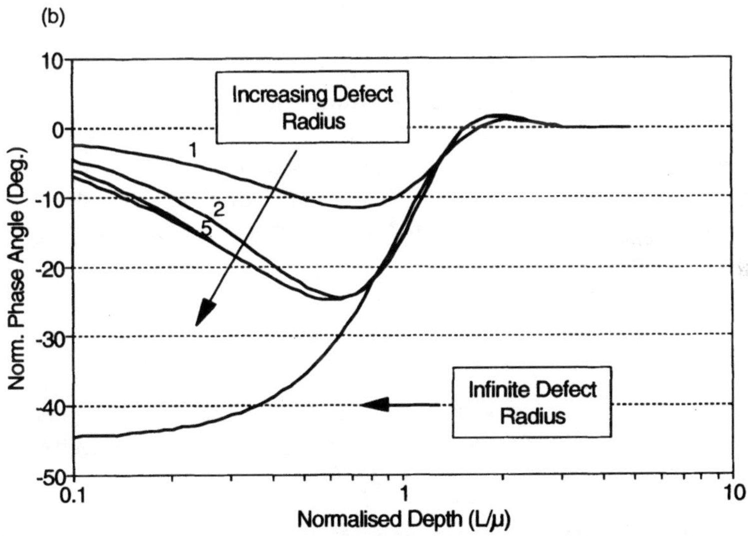

Figure 2.15 *The amplitude (a) and phase (b) of the signal over the center of disk defects, radii 1, 2, 5 and 10 mm, shown as a function of the normalized defect depth.*

if it is assumed that the temperature across the surface of the defect is uniform.

This is a good assumption for a thermal diffusion length away from the edge of the defect. A further approximation in the expression is that only a single reflected term needs to be used in computing the surface temperature profile. A more precise calculation would consider multiple reflections between the defect and sample surface. The factor of two preceding the second term in the above equation accounts for the fact that the point heat sources on the defect surface can only flow towards the sample surface. This is due to the requirement that the defect surface is insulating. The expression must be regarded as a first-order approximation. It is used here to illustrate the principal effects caused by the finite extent of real thermal wave reflectors.

In Fig. 2.15 the amplitude and phase of the thermal wave signals at the surface over the centers of the disks of different radii are plotted as a function of normalized depth (defect depth/thermal diffusion length). When the radius is large compared with the thermal diffusion length the one-dimensional interference results (Figs 2.7 and 2.8) for R_b equal to unity are recovered. For smaller defects the magnitudes of these effects diminish as their effective reflecting power reduces with their area. Other effects caused by the finite size of defects, probes or detectors will be discussed in later chapters.

References

Bennett, C.A. and Patty, R.R., 'Thermal wave interferometry: a potential application of the photoacoustic effect', *Appl. Opt.*, **21** (1982) 49–54.

Further reading

Alzofan, F.E., 'Retardation and diffraction aspects of the conduction of heat in solids', *Am. J. Phys.*, **30** (1962) 285–293.

Carslaw, H.S. and Jaeger, J.C., *Conduction of heat in solids*, Oxford University Press (1959).

Green, G., 'Some problems in heat conduction', *Phil. Mag.*, **3** (1927) 784–800.

Incropera, F.C. and DeWitt, D.C., *Fundamentals of heat and mass transfer*, John Wiley, London, (1985).

Lepoutre, F., Fournier, D. and Boccara, A.C., 'Non-destructive control of welding using the mirage deflection', *J. Appl. Phys.*, **57** (1985) 1009–15.

McDonald, F.A., 'Photoacoustic effect and the physics of waves', *Am. J. Phys.*, **48** (1980) 41–47.

Patel, P.M., Almond, D.P. and Reiter, H., 'Thermal wave detection and characterisation of subsurface defects', *Appl. Phys.*, **B43** (1987) 9–15.

Rosencwaig, A. and Gersho, A., 'Theory of the photoacoustic effect in solids', *J. Appl. Phys.*, **47** (1976) 64–69.

Wetsel, G.C. and McDonald, F.A., 'Subsurface structure determination using photothermal laser beam deflection', *Appl. Phys. Lett.*, **41** (1982) 926–28.

3

Optical generation of thermal waves – photothermal effect

This book covers a wide range of topics, techniques and phenomena which are linked by their dependence on the photothermal effect. By far the most common way of generating thermal waves is to heat a sample with a modulated laser beam. This method relies on a fraction of the optical energy being absorbed by the sample and being converted rapidly to heat. The great advantages of the photothermal effect are that it is intrinsically non-contactive and that it can provide the means of producing a very precisely defined region of modulated heating. Additionally, spectroscopic information can be obtained about some materials through their selective absorption of parts of the electromagnetic spectrum. For all applications, the net photothermal effect will depend on the optical properties of the material under investigation as well as its thermal properties. In this chapter we introduce the basic optical properties and show how they influence the photothermal effects produced in different types of material.

3.1 Surface reflectivity

Reflectivity is the optical parameter which determines the fraction of incident optical radiation that is reflected from the surface of a sample and, consequently, the remaining fraction that enters a sample, where it may contribute to sample heating. In this section we outline the interaction of an electromagnetic wave with a medium, and introduce the materials' properties that characterize electromagnetic wave propagation in a medium.

When electromagnetic radiation is incident on a sample it will, in general, be partly reflected and partly absorbed or transmitted,

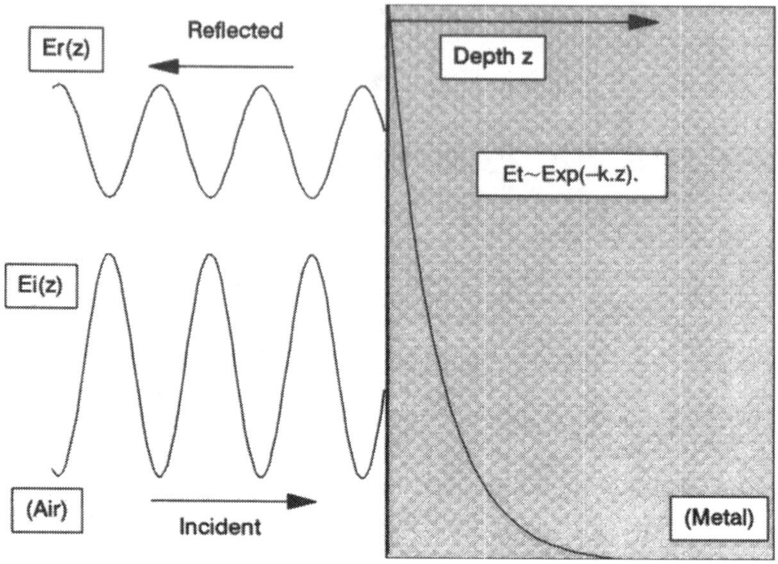

Figure 3.1 *Reflection and decay of electromagnetic fields at the surface of a metal.*

as shown schematically in Fig. 3.1. It is sufficient here to consider the electric field component of an electromagnetic wave. The propagation of the incident electric field component is given by the expression:

$$E(z) = E(o)\exp\left(\frac{-2\pi n''z}{\lambda} + \mathrm{j}\frac{2\pi n'z}{\lambda} + \mathrm{j}\omega t\right)$$
$$E(z) = E(o)\exp\left(\mathrm{j}\frac{2\pi Nz}{\lambda} + \mathrm{j}\omega t\right) \tag{3.1}$$

where $E(o)$ is the surface amplitude of the electric field and the surface is the plane $z = 0$. The coefficients n' and n'' are combined to define the complex refractive index of the material $N = n' + \mathrm{j}n''$, where n'' is the attenuation, n' the refractive index and λ is the wavelength of the electromagnetic wave. The decay of optical intensity is obtained using the Poynting theorem:

$$I(z) = N|E(z)|^2 = I(o)\exp(-\mathrm{j}4\pi Nz/\lambda)$$
$$I(z) = I(o)\exp(-4\pi n''z/\lambda + \mathrm{j}4\pi n'z/\lambda) \tag{3.2}$$

Table 3.1 A comparison between the optical absorption $1/\beta$ and thermal diffusion length μ of various materials

Material	λ (μm)	n'	n''	$R(\lambda)$ (%)	β (μm^{-1})	$1/\beta$ (μm)	μ at 100 Hz (μm)
Fe	1.0	3.20	4.10	62	51.5	0.0194	269
Ni	1.0	2.85	5.10	72	64.1	0.0156	270
W	1.0	3.00	3.50	57	43.9	0.0227	466
Ti	1.0	3.30	3.35	55	42.1	0.0237	172
Ge	1.0	4.44	1.23	40	15.4	0.065	332
Si	1.0	3.5	0.00018	30	0.0022	454	532
Glass	1.0	1.5	≈0	4	≈0		52
Polystyrene	1.0	1.59	≈0	5.3	≈0		14

and the surface reflectance and transmittance can be written as:

$$R(\lambda)=\left|\frac{1-N(\lambda)}{1+N(\lambda)}\right|^2 \qquad T(\lambda)=\left|\frac{2N(\lambda)}{1+N(\lambda)}\right|^2 \tag{3.3}$$

The incident optical intensity decays exponentially in a conducting medium, such as a metal, with an absorption coefficient given by $\beta = 4\pi n''/\lambda$. The inverse of this quantity, the optical absorption depth or skin depth, is the distance over which the incident light field decays to 1/e of its surface amplitude. For non-absorbing materials the coefficient n'' is zero. For metallic materials the optical 'skin depth' is a few tens of nanometers. In semiconductors this depth can extend to micrometers and for insulators to millimeters. Table 3.1 compares the optical absorption depths for metals, semiconductors and insulators with values of the thermal diffusion length, introduced in Chapter 2. The latter is seen to be much larger than the optical penetration depth of metallic materials. This is the justification for the assumption that thermal waves generated by the photothermal effect in metals can be taken to be generated in the surface plane.

The coefficients n' and n'', which are determined by the electronic energy band structure of a material, can be obtained experimentally from measurements of the complex dielectric constant:

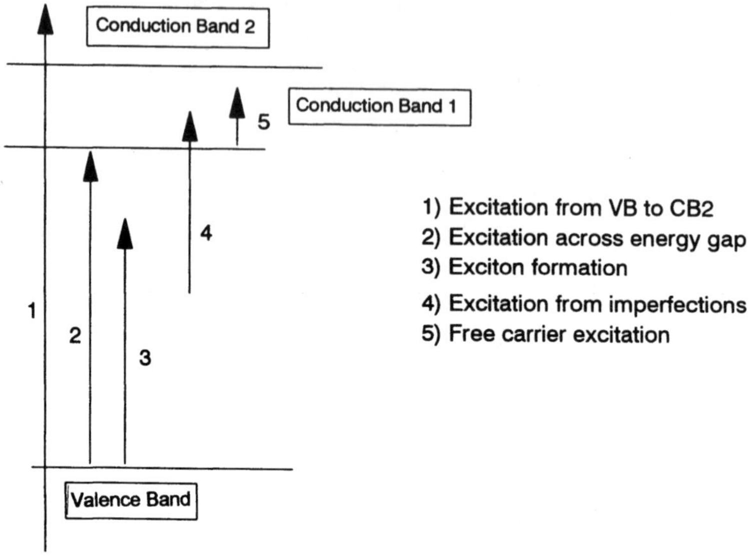

Figure 3.2 *Schematic illustration of various inter- and intraband transitions in a material.*

$$\varepsilon(\omega) = N^2 = \varepsilon'(\omega) + \mathrm{j}\varepsilon''(\omega) \tag{3.4}$$

$$n' = \left[\frac{(\varepsilon'^2 + \varepsilon''^2)^{1/2} + \varepsilon'}{2}\right]^{1/2}, \; n'' = \left[\frac{(\varepsilon'^2 + \varepsilon''^2)^{1/2} - \varepsilon'}{2}\right]^{1/2} \tag{3.5}$$

Up to mid-infrared frequencies three components can be identified which contribute to the total dielectric function of a material:

$$\varepsilon(\omega) = \varepsilon_{\mathrm{L}}(\omega) + \varepsilon_{\mathrm{B}}(\omega) + \varepsilon_{\mathrm{F}}(\omega) \tag{3.6}$$

where $\varepsilon_{\mathrm{L}}(\omega)$ describes effect of the absorption of incident radiation by the lattice, $\varepsilon_{\mathrm{B}}(\omega)$ the absorption of radiation by interband and $\varepsilon_{\mathrm{F}}(\omega)$ intraband transitions. These processes are illustrated schematically in Fig. 3.2. In metallic materials the absorption of electromagnetic radiation is mostly by intraband transitions where the free conduction electrons are excited to higher energy states within the conduction band (process 5 in Fig. 3.2). The absorption of radiation by this mechanism is known as free carrier absorption.

The free electron gas, Drude model, expression for the complex dielectric function is:

$$\varepsilon_{\mathrm{F}}(\omega) = \varepsilon'(\omega) + \mathrm{j}\varepsilon''(\omega) \tag{3.7}$$

where

$$\begin{aligned} \varepsilon'(\omega) &= 1 - \frac{\omega_{\mathrm{p}}^2\tau^2}{1+\omega^2\tau^2} \\ \varepsilon''(\omega) &= \frac{\omega_{\mathrm{p}}^2\tau}{\omega(1+\omega^2\tau^2)} \\ \omega_{\mathrm{p}} &= \left(\frac{N_{\mathrm{f}}e^2}{\varepsilon_0 m}\right)^{1/2} \end{aligned} \tag{3.8}$$

and N_{f} is the number of free electrons, τ is the electron collision relaxation time, m and e the electronic mass and charge, ε_0 is the permittivity of free space and ω_{p} is the plasma frequency (natural frequency of collective motion of charges). In metals this frequency usually lies in the ultraviolet spectral region which accounts for the

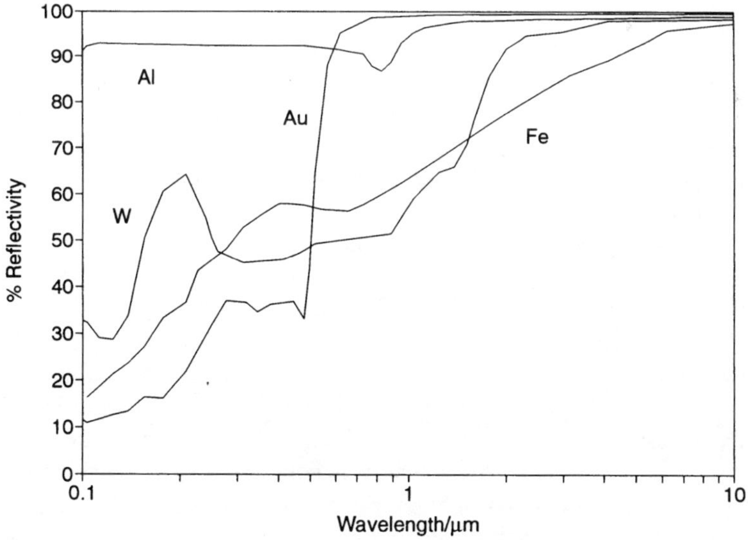

Figure 3.3 *Reflectivity of metals (Al – aluminum, Au – gold, Fe – iron and W – tungsten) with wavelength.*

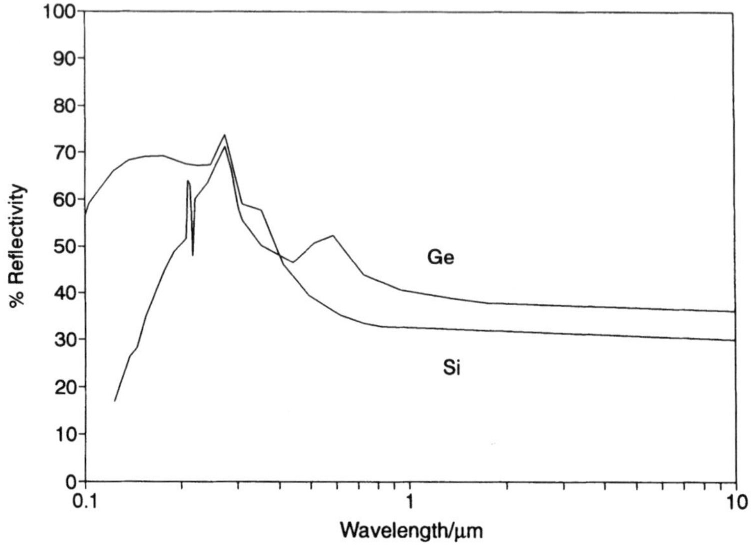

Figure 3.4 *Reflectivity of silicon and germanium single crystal.*

ultraviolet transparency of metals. In this spectral region the refractive index of the metal is real and its absorption index is small and real. At much lower frequencies (mid-infrared wavelengths and higher) the absorption index increases leading to high absorption in the infrared spectral region and high reflectivity. In this spectral region the surface reflectivity is given by the Hagen–Rubens expression which relates reflectivity to electrical conductivity:

$$R \approx 1 - \left(\frac{8\omega\varepsilon_0}{\sigma_0} \right)^{1/2} \tag{3.9}$$

where σ_0 is the dc electrical conductivity. This predicted variation can be seen in Fig. 3.3 where the reflectivities of some metals are shown plotted against wavelength. If the incident radiation is such that interband transitions are possible an increase in both optical absorption and surface reflectivity occurs. This is the case in gold and gives the metal its characteristic yellow colour.

In semiconducting materials the optical reflectivity shows a strong variation with wavelength near the band-gap energy (processes 1, 2 and 4). Semiconducting materials are generally opaque in the visible and translucent in the infrared. In the visible spectral region free carrier generation may occur, leading to an increased absorption and reflectivity as shown in Fig. 3.4. The free carrier contribution is much smaller than in metals because the number of free carriers is so much lower. An important difference between metals and semiconductors is that the free carrier concentration of semiconducting materials can be increased significantly by increasing the sample temperature or by doping. At frequencies lower than that of the fundamental for absorption (below the band-gap energy) the optical properties of semiconductors are similar to those of a dielectric material. Dielectric materials have a reflectivity variation which is almost independent of wavelength. Absorption in these materials is negligible and they are typically transparent in the visible and infrared spectral regions. At very high frequencies (high photon energies) excitations by fundamental absorption processes occur, leading to high absorption and surface reflectivity. This is illustrated in Fig. 3.5 which shows the reflectivity variation of natural diamond. In homo-polar materials lattice absorption via photon–phonon coupling leads to increased absorption in the mid- and far-infrared spectral regions.

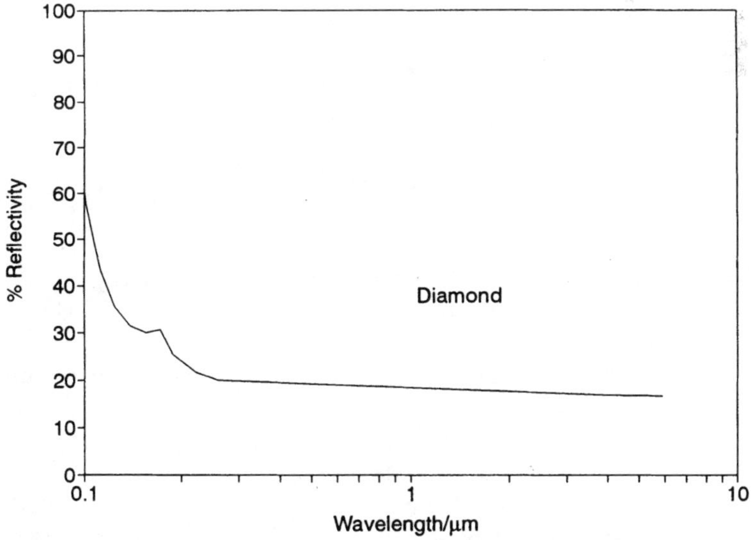

Figure 3.5 *Reflectivity of diamond.*

Photothermal studies rely on a significant proportion of incident optical energy being absorbed and converted to heat. A combination of high absorption and low surface reflectivity are required for efficient heat generation. Electrically conducting materials are characterized by high reflectivity and consequently only a small fraction of any incident optical energy is converted to heat. It may be possible to improve matters by selecting an optical source for a particular material. For metallic materials short wavelength ultraviolet radiation should be more effective than visible light as the surface reflectivity decreases in the ultraviolet whilst absorption remains high enough to produce heating. In practice, the use of ultraviolet laser photothermal sources is rare and most work is performed using high power visible lasers and very sensitive detection systems to make up for the low photothermal efficiency.

3.2 Bulk absorption

Many non-metallic materials absorb optical energy and convert it to heat. The optical absorption processes in these materials involve atomic or molecular transitions rather than free conduction electrons. In general, electromagnetic waves penetrate deeper into non-metallic materials (their optical absorption depths are longer) and it is often impossible to consider the photothermal effect to provide a simple surface heat source.

The heat generated in an element of thickness dx at a depth x in a semi-infinite optically absorbing medium is given by (Fig. 3.6):

$$\Delta Q(x) = -\eta \frac{\mathrm{d}I(x)}{\mathrm{d}x}\mathrm{d}x = \eta[1 - R(\lambda)]I_0\beta\exp(-\beta x)\mathrm{d}x \tag{3.10}$$

assuming an exponential attenuation of the optical energy entering the medium and a surface reflectivity $R(\lambda)$. β is the optical absorption coefficient and η_1 is the light to heat conversion efficiency of the sample. The latter factor is usually taken to be unity for solid materials, as is assumed below. The magnitudes of elemental heat sources such as this are seen to decay in amplitude exponentially with depth x. In a photothermal experiment, thermal waves originating from these elemental sources will have amplitudes depending on both the optical damping factor β and the thermal wave damping term involving the thermal diffusion length μ. The

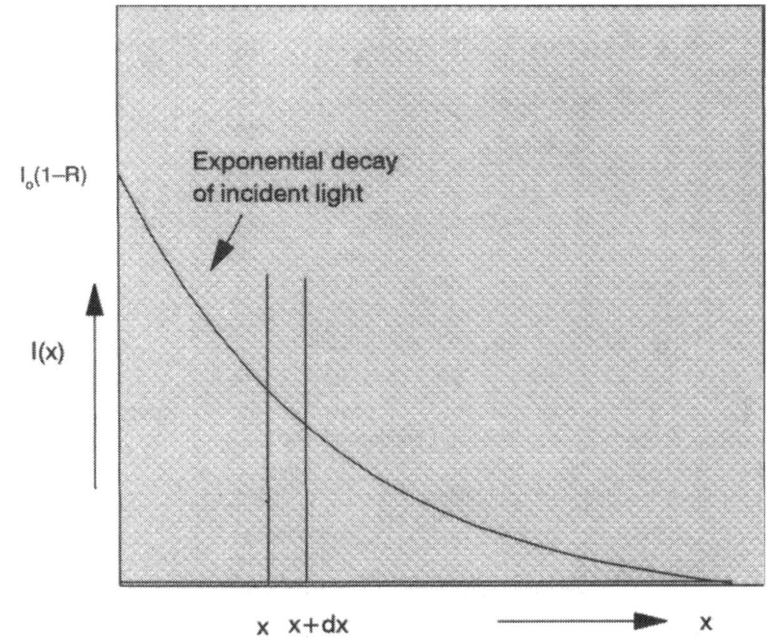

Figure 3.6 *Schematic illustration of the location and amplitude of an optical heat generation source in a semi-infinite translucent medium.*

resulting modulation in sample surface temperature is obtained by a summation of the contributions from elemental sources at all depths:

$$T(\omega) = [1 - R(\lambda)]I_0 \beta T_g \int_0^{\infty} \frac{\exp[-(\beta + \sigma)x]}{4k\sigma} \mathrm{d}x \ \exp(\mathrm{j}\omega t) \tag{3.11a}$$

$$T(\omega) = \frac{[1 - R(\lambda)]I_o T_g}{4k\sigma} \left(\frac{\beta}{\beta + \sigma} \right) \ \exp(\mathrm{j}\omega t) \tag{3.11b}$$

T_g is the thermal wave transmission coefficient at the sample surface. This equation reduces to that in which heat is generated at the sample surface (equation (2.22) for $x = 0$ and Q_0 defined here photothermally) for a strongly absorbing medium in which $\beta \gg \sigma$.

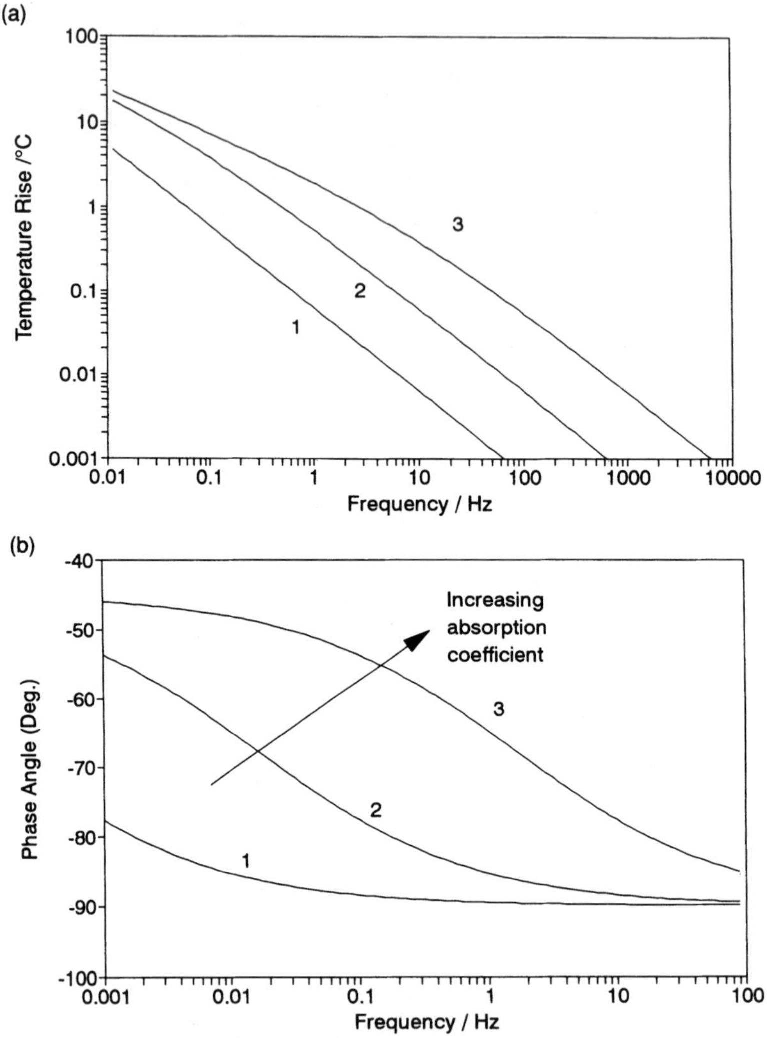

Figure 3.7 *Temperature (a) and phase (b) variation with frequency for an absorbing semi-infinite medium. Optical absorption coefficients: (1)*10^2*, (2)*10^3 *and (3)* 10^4 m^{-1}.

For the opposite case where $\beta \ll \sigma$ and the sample is translucent the above equation reduces to:

$$T(\omega) = \frac{[1 - R(\lambda)] I_0 T_g \beta}{4 j \rho c \omega} \quad \exp(j\omega t) \tag{3.12}$$

The complex surface temperature decays with the inverse of the modulation frequency and the thermal capacity whilst it is directly proportional to the sample's optical absorption coefficient. The phase of the surface temperature is independent of frequency and equals –90°. In general, the conditions will fall between these two extremes and depend on modulation frequency because the thermal wave number σ is frequency dependent. Fig. 3.7 shows the amplitude and phase variation of $T(\omega)$, equation (3.11), with the frequency for several values of optical absorption coefficient, illustrating the transition from photothermal opaqueness to photothermal transparency.

3.3 Absorption in a thin surface layer

When absorption occurs in a thin film or a coating which lies on a non-absorbing semi-infinite substrate, the surface temperature due to optical absorption can be obtained by the method of wave trains (thermal wave interference) presented in Chapter 2. Instead of a surface heat generation source, the heat source is distributed through the thickness of the thin film. The surface temperature is obtained by summing up the wave trains arising from the heat sources throughout the film, taking account of reflections from the film–substrate interface and the film–surface (gas) interface. Let the suffixes 0, 1 and 2 refer to the gas, film and substrate physical properties. From equation (3.11a), thermal waves generated in a unit area slice of the coating material with thickness dx lying at a distance x beneath the surface will have a strength A given by:

$$A = \frac{[1 - R(\lambda)]\beta_1 I_o}{4k_1\sigma_1} \exp(-\beta_1 x)\,\mathrm{d}x \tag{3.13}$$

For waves reflected from the film–air boundary the contribution is:

$$\begin{aligned} &AT_{01}\left[e^{-\sigma_1 x} + R_{21}R_{01}e^{-\sigma_1(2L_1+x)} + \ldots + (R_{21}R_{01})^n e^{-\sigma_1(2nL_1+x)}\right] \\ &\quad = AT_{01}e^{-\sigma_1 x}\sum_{n=0}^{n=\infty}(R_{21}R_{01})^n e^{-2n\sigma_1 L_1} \\ &\quad = \frac{AT_{01}e^{-\sigma_1 x}}{1 - R_{21}R_{01}e^{-2\sigma_1 L_1}} \end{aligned} \tag{3.14}$$

in which L_l is the coating thickness.

For waves reflected from the film–substrate boundary their contribution is:

$$\begin{aligned} &AT_{01}\left[R_{21}\mathrm{e}^{-\sigma_1(2L_1-x)}+R_{21}(R_{21}R_{01})\mathrm{e}^{-\sigma_1(4L_1-x)}+\ldots\right.\\ &\quad\left.+R_{21}(R_{21}R_{01})^n\mathrm{e}^{-\sigma_1(2(n+1)L_1-x)}\right]\\ &=AT_{01}R_{21}\mathrm{e}^{-\sigma_1(2L_1-x)}\sum_{n=0}^{n=\infty}(R_{21}R_{01})^n\mathrm{e}^{-2n\sigma_1L_1}\\ &=\frac{AT_{01}R_{21}\mathrm{e}^{-\sigma_1(2L_1-x)}}{1-R_{21}R_{01}\mathrm{e}^{-2\sigma_1L_1}} \end{aligned} \tag{3.15}$$

R_{21} and R_{01} are the thermal wave reflection coefficients at the film–substrate and film–air interfaces, respectively, and T_{01} is the film–air thermal wave transmission coefficient. The value of these coefficients can be obtained from their definitions in Chapter 2:

$$\begin{aligned} R_{01}&=\frac{(1-g)}{(1+g)},\quad T_{01}=\frac{2}{1+g},\quad g=\sqrt{\frac{(\rho ck)_0}{(\rho ck)_1}}\\ R_{21}&=\frac{(1-b)}{(1+b)},\qquad\qquad b=\sqrt{\frac{(\rho ck)_2}{(\rho ck)_1}} \end{aligned} \tag{3.16}$$

Summing up the two contributions and substituting for A we obtain:

$$\frac{[1-R(\lambda)]\beta_1I_0T_{01}}{4k_1\sigma_1\left(1-R_{21}R_{01}\mathrm{e}^{-2\sigma_1L_1}\right)}\left[\mathrm{e}^{-(\beta_1+\sigma_1)x}+R_{21}\mathrm{e}^{-2\sigma_1L_1}\mathrm{e}^{-(\beta_1-\sigma_1)x}\right]\mathrm{d}x \tag{3.17}$$

Integrating over the film thickness results in the complex surface temperature variation at x=0:

$$\frac{[1-R(\lambda)]\beta_1I_0T_{01}}{4k_1\sigma_1\left(1-R_{21}R_{01}\mathrm{e}^{-2\sigma_1L_1}\right)}\int_0^{L_1}\left[\mathrm{e}^{-(\beta_1+\sigma_1)x}+R_{21}\mathrm{e}^{-2\sigma_1L_1}\mathrm{e}^{-(\beta_1-\sigma_1)x}\right]\mathrm{d}x \tag{3.18}$$

$$T(\omega,0)=\frac{[1-R(\lambda)]\beta_1I_0T_{01}}{4k_1\sigma_1\left(1-R_{21}R_{01}\mathrm{e}^{-2\sigma_1L_1}\right)}\times$$
$$\left(\frac{1-\mathrm{e}^{-(\beta_1+\sigma_1)L_1}}{\beta_1+\sigma_1}+R_{21}\mathrm{e}^{-2\sigma_1L_1}\,\frac{1-\mathrm{e}^{-(\beta_1-\sigma_1)L_1}}{\beta_1-\sigma_1}\right)\mathrm{e}^{\mathrm{j}\omega t} \tag{3.19}$$

This equation provides a complete description of photothermal wave generation in a thin film. The effect of subsurface optical heat generation is to introduce two factors $A_{\pm}$ to expression (2.38) for thermal wave interference in an opaque thin film in which surface heating alone occurs:

$$A_{\pm} = \frac{1 - e^{-(\beta \pm \sigma)L}}{\beta \pm \sigma} \tag{3.20}$$

When the film is highly absorbing these two terms tend to $1/\beta$ and the final surface temperature expression is independent of optical absorption coefficient. For a film with a very low absorption coefficient these $A_{\pm}$ terms become independent of β and the sample surface temperature is left directly dependent on the absorption coefficient, as is evident from equation (3.19).

The surface temperature expression (3.19) reduces to simpler forms for various limiting conditions. Fig. 3.8 illustrates schematically some of these conditions which are discussed below. In this figure μ_1 is the thermal diffusion length and $\mu_\beta = 1/\beta$, the optical skin depth of the surface layer. When the surface layer can be regarded as being optically opaque, $\mu_\beta \ll L_1$, equation (3.19) simplifies to:

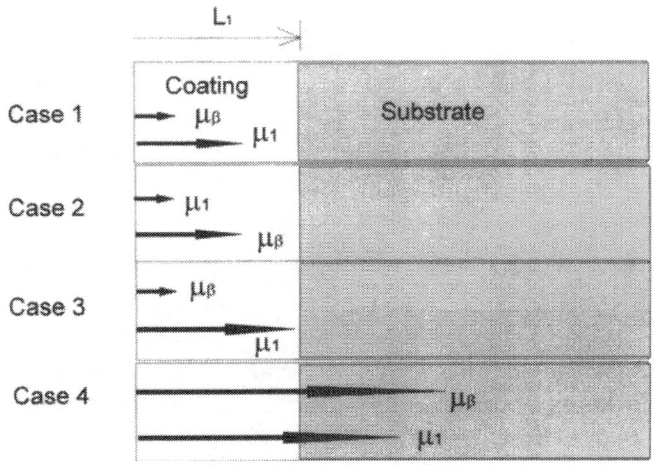

Figure 3.8 *Schematic diagram showing a set of special conditions for the solution of equation (3.19).*

$$T(\omega,0)=\frac{[1-R(\lambda)]\beta_1 I_0 T_{01}}{4k_1\sigma_1\left(1-R_{21}R_{01}e^{-2\sigma_1 L_1}\right)}\left(\frac{1}{\beta_1+\sigma_1}+\frac{R_{21}e^{-2\sigma_1 L_1}}{\beta_1-\sigma_1}\right)e^{j\omega t} \tag{3.21}$$

This expression can be simplified further in three special cases determined by the relative sizes of μ_1, the sample thickness, L_1, and the optical absorption depth μ_β.

Case 1: $\mu_1 \ll L_1$ and $\mu_\beta \ll \mu_1$

This is the case when thermal diffusion length in the sample is much smaller than the layer thickness, but greater than the optical penetration depth. The sample surface temperature becomes:

$$T(\omega,0)=\frac{[1-R(\lambda)]I_0}{2k_1\sigma_1}\exp(j\omega t) \tag{3.22}$$

Under this set of conditions the surface temperature depends only on the thermal properties of the layer and its optical reflectivity. The surface temperature amplitude decays as $f^{1/2}$ ($\sigma_1 \propto \sqrt{f}$, f=modulation frequency). The phase of the surface temperature lags the applied heat source by 45° because $\sigma_1 \propto (1 + j)$, and is independent of the optical properties of the layer. These are the standard characteristics of an opaque thermally thick layer.

Case 2: $\mu_1 \ll L_1$ and $\mu_\beta > \mu_1$

In this case the layer is also thermally thick but it is not opaque as the optical penetration depth is larger than thermal diffusion length. The surface temperature is given by:

$$T(\omega,0)=\frac{[1-R(\lambda)]I_0}{2k_1\sigma_1}\frac{\beta_1}{\beta_1+\sigma_1}\exp(j\omega t) \tag{3.23}$$

We note that in addition to the dependence of the surface temperature on the optical reflectivity there is an additional dependence on the optical absorption coefficient β_1. The phase of the surface temperature can be obtained from the real and imaginary parts of the complex surface temperature:

$$\phi = \tan^{-1}\left(\frac{\mathrm{Im}(T)}{\mathrm{Re}(T)}\right) = -\tan^{-1}\left(1 + \frac{2}{\beta_1 \mu_1}\right) \tag{3.24}$$

The phase depends only on parameter $\beta_1\mu_1$. This relationship could be used to determine the bulk optical absorption coefficient providing the thermal diffusivity of the layer was known. Equation (3.24) also shows that for sufficiently small values of the optical absorption coefficient the phase angle becomes independent of this parameter and approaches –90°. In this limit the surface temperature is given by

$$T(\omega,0) = \frac{[1 - R(\lambda)]\beta_1 I_0}{2k_1\sigma_1^{\,2}} \exp(\mathrm{j}\omega t) \tag{3.25}$$

Now the surface temperature amplitude is directly proportional to the β_1 and decays inversely with the modulation frequency of the heat source and the thermal capacity ρc.

Case 3: $\mu_1 \approx L_1$ and $\mu_\beta > \mu_1$

In this limit the thermal diffusion length is comparable to the layer thickness and the heat deposition region is much narrower than diffusion length (surface heating). The surface temperature is now given by equation (2.38) with A defined photothermally:

$$T(\omega,0) = \frac{[1 - R(\lambda) I_0]}{2k_1\sigma_1}\left(\frac{1 + R_{21}\mathrm{e}^{-2\sigma_1 L_1}}{1 - R_{21}\mathrm{e}^{-2\sigma_1 L_1}}\right) \exp(\mathrm{j}\omega t) \tag{3.26}$$

where R_{21} is the thermal wave reflection coefficient between medium 2 and medium 1. As discussed previously the surface temperature amplitude depends on the layer thickness, the thermal properties of the layer and the backing, and the optical reflectivity of the surface layer. The phase of the temperature is independent of optical reflectivity, depending only on the thermal properties of the layer and backing medium and the thickness of the surface layer. If the surface layer is thermally thin, $\mu_1 \gg L_1$, then the surface temperature is determined by the thermal impedance of the backing layer:

$$T(\omega,0)=\frac{[1-R(\lambda)]I_0}{2k_2\sigma_2}\exp(j\omega t) \tag{3.27}$$

owing to the negligible temperature difference over the layer thickness. The phase becomes –45°.

When the surface layer is optically thin $\mu_\beta \gg L_1$, equation (3.19) simplifies to:

$$T(\omega,0)=\frac{[1-R(\lambda)]\beta_1 I_0 T_{01}}{4k_1\sigma_1\left[1-R_{21}R_{01}e^{-2\sigma_1 L_1}\right]}\times$$

$$\left(\frac{1-e^{-\sigma_1 L_1}}{\beta_1+\sigma_1}+R_{21}e^{-2\sigma_1 L_1}\frac{1-e^{\sigma_1 L_1}}{\beta_1-\sigma_1}\right)\exp(j\omega t) \tag{3.28}$$

Case 4: $\mu_1 \gg L_1$

In this limit the surface layer is, in addition to being optically thin, also thermally thin. The surface temperature is now given by:

$$T(\omega,0)=\frac{[1-R(\lambda)]I_0\beta_1 L_1}{2k_2\sigma_2{}^2}\exp(j\omega t) \tag{3.29}$$

The magnitude of the surface temperature is determined by the optical thickness, $\beta_1 L_1$ and its phase is –90°. This situation is fulfilled by thin surface layers with low optical absorption coefficients.

Figures 3.9(a) and (b) show the thermal amplitude and phase respectively as a function of modulation frequency for a translucent medium (calculated using the thermal properties of the translucent ceramic yittria stabilized zirconia) on a non-absorbing substrate. The layer thickness is 0.25 mm and several values of the optical absorption coefficient are considered. The surface temperature is seen to decrease with increasing frequency. Thermal interference effects occur in the middle part of the graph where there is a small enhancement of the surface temperature. The limiting conditions discussed above can also be identified in these figures. The layer is thermally thin at low frequencies and becomes thermally thick at the highest frequencies. In the phase angle plots of Fig. 3.9(b) these limiting and intermediate cases are clearly evident. At low frequencies the phase is close to –45°. At high

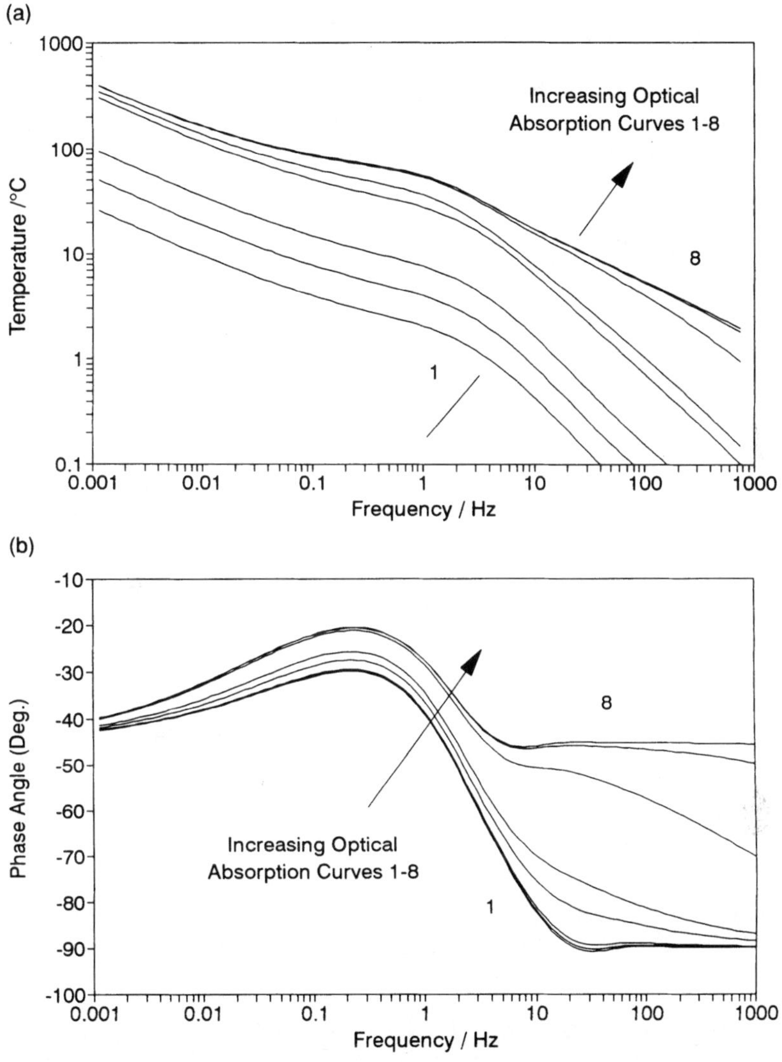

Figure 3.9 *The surface temperature (a) and phase (b) variation with modulation frequency for a coating substrate system, the coating having optical absorption coefficients: (1) 250, (2) 500, (3) 1000, (4) 5000, (5) 1×10^4, (6) 1×10^5, (7) 1×10^6 and (8) 1×10^7 m^{-1}. The thermal properties of the coating, thickness 0.25 mm (yttria stabilized zirconia), $\rho = 5000$, $c = 500$, $k = 1$; substrate (steel), $\rho = 8230$, $c = 480$, $k = 13.4$ SI units.*

frequencies the phase approaches –90° in the low optical absorption coefficient samples whilst in the high optical absorption coefficient samples it returns to –45°. At intermediate frequencies the phase rises above –45° to a peak and at higher frequencies and with increasing translucency it drops below –45°.

3.4 Layer transparency

The above analysis neglected any contribution to heat generation caused by the absorption of light in the substrate. Such a heat source term may be important when considering translucent or transparent surface layers. Another potential correction term arises from taking account of optical reflection at the layer–substrate and layer–air interfaces. These effects cause optical interference in sufficiently thin coating layers. The problem of substrate absorption will be considered first, neglecting optical reflections at the two interfaces. There are, in this case, two distributed heat source terms: one in the surface layer (given by equation (3.10)) and another located in the substrate. The latter corresponds to an elemental thermal source term $Q_2(x)$ at a depth x below the substrate surface:

$$Q_2(x) = \eta_2[1 - R_1(\lambda)]I_0\beta_2 e^{-(\beta_1+\sigma_1)L_1} e^{\beta_2 x} \tag{3.30}$$

where the subscript 2 refers to the properties of the substrate.

The solution of the heat diffusion equation for the layered sample can be written as a sum of two components T_1 and T_2. The first term T_1 is that due to heating the surface layer and the second, T_2, that due to heat generation in the substrate. T_1 is given by equation (3.19).

$$T_2 = \frac{\eta_2[1 - R(\lambda)]I_0\beta_2 e^{-(\beta_1+\sigma_1)L_1}}{[k_1\sigma_1 + k_2\sigma_2][\beta_2 + \sigma_2][1 - R_{21}e^{-2\sigma_1 L_1}]}\exp(j\omega t) \tag{3.31}$$

Inspection of T_2 shows that it reduces to zero when either β_2 is zero or when the surface layer is optically opaque or thermally thick. Figure 3.10 presents a plot of the complex surface temperature with and without inclusion of substrate absorption for a coating layer with an optical absorption of 1000 m^{-1}. The effect of substrate absorption on surface temperature increases as modulation fre-

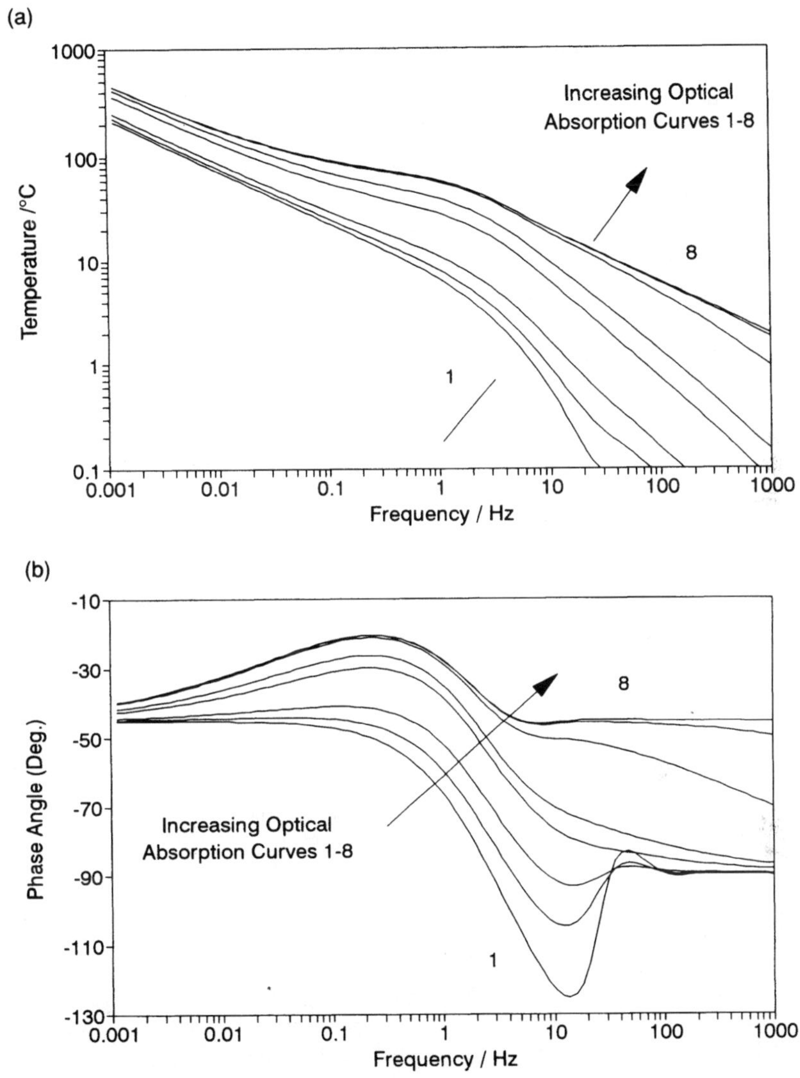

Figure 3.10 *The surface temperature (a) and phase (b) variation with frequency for the coating substrate system, Fig. 3.9 with coating optical coefficient of 1000* m^{-1}*, taking into account substrate absorption. Layer thermal properties are as those given in Fig. 3.9. Substrate optical absorption coefficients: (1) 250, (2) 500, (3) 1000, (4) 5000, (5)* 1×10^4*, (6)*1×10^5*, (7)* 1×10^6 *and (8)* 1×10^7 m^{-1}*.*

(a)

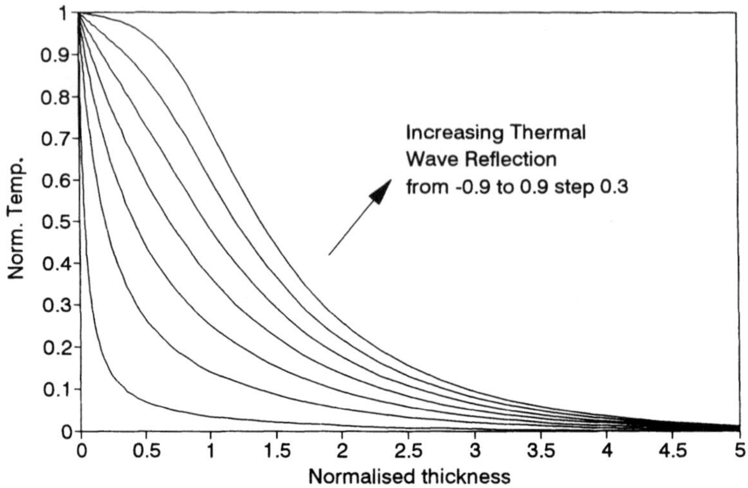

(b)

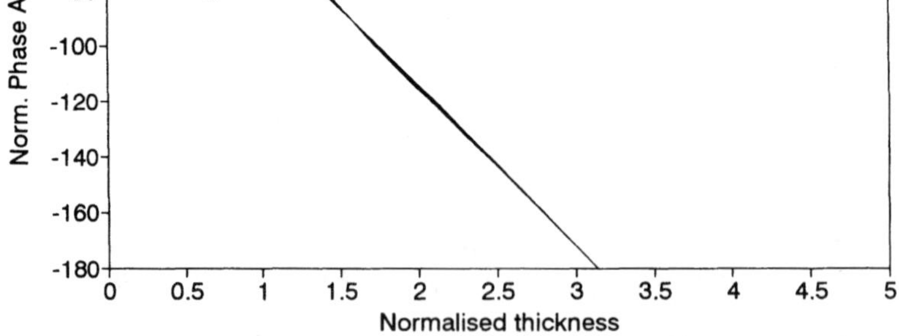

Figure 3.11 *A normalized surface temperature (a) and phase (b) versus thermal thickness* L/μ_l *for a transparent surface layer over an optically opaque substrate shown for thermal wave reflection coefficients ranging from –0.9 to 0.9 in steps of 0.3.*

quency is reduced. At higher frequencies the contribution from substrate absorption reduces rapidly and the surface temperature becomes determined by T_1 alone. These effects are caused by the decrease in thermal diffusion length with frequency. Thermal waves generated in the substrate have to pass through the coating to reach the surface to affect the surface temperature. The attenuation of these waves by the coating increases with modulation frequency, explaining the decrease in the significance of the substrate term with increasing frequency. The effect of substrate absorption on the phase–frequency dependence is to produce a locus which is lowered in magnitude with respect to the 'no substrate' absorption case.

When the surface layer is transparent, $\beta_1 = 0$, no optical absorption or thermal generation occurs in the layer. If the substrate is absorbing and optically opaque, the surface temperature (neglecting optical interference) is then determined by T_2 alone:

$$T(\omega,0) = \frac{\eta I(\bar{\lambda})}{(k_1\sigma_1 + k_2\sigma_2)} \frac{e^{-\sigma_1 L_1}}{(1 - R_{21}e^{-2\sigma_1 L_1})} \exp(j\omega t) \tag{3.32}$$

where η is a factor which accounts for the light intensity reflection losses and the light–heat conversion efficiency and $I(\bar{\lambda})$ an effective light intensity term. This equation should be compared with equation (3.26), the thermal wave interference equation for an opaque surface layer. In the transparent layer thermal wave interference also occurs, but as the heat source is located at the substrate surface, the surface temperature is exponentially damped with increasing layer thickness. In Fig. 3.11 thermal amplitude and phase are plotted, normalized with respect to the substrate. Compared with the $R = 0$ curve ($b_{21} = 1$ thermally matched surface and substrate) constructive interference occurs for $b_{21} > 1$ and destructive interference for $b_{21} < 1$. The phase angle variation also shows thermal wave interference phenomena but its overall change (max. to min.) is much reduced when compared with the optically opaque case (compare with Figure 2.8) and the phase variation at high thermal thickness is independent of the thermal wave reflection coefficient. One potential benefit of this is the possibility of extending the thickness detection capability above 1.5 thermal diffusion lengths. The upper limit is set by the signal-to-noise ratio of the detection system.

3.5 Light interference effects

In the above analysis it was assumed that no internal optical reflections occurred in the layer. This assumption is invalid in low optical absorption layers in which the reflected component within the layer may be comparable in magnitude to the first component transmitted through the layer. To solve the heat diffusion equation for this case the source term of equation (3.10) will need to be modified to account for the light intensity variation in the surface layer. For an absorbing layer on a transparent substrate this source term is:

$$Q(x)=\frac{\eta|t_1|^2\beta_1 I_0(e^{\beta_1 x}+|r_2|^2 e^{-\beta_1(2L_1+x)})}{2\left[1+|r_1|^2|r_2|^2 e^{-2\beta_1 L_1}-2r_1r_2e^{-\beta_1 L_1}\cos(2\delta L_1)\right]} \tag{3.33}$$

where t_1, r_1 are the optical transmission and reflection coefficients at the air–layer interface and r_2 the reflection coefficient at the layer–substrate interface and $\delta = 2\pi n/\lambda$. Solving the heat diffusion equation results in the following expression for the sample surface temperature:

$$T(\omega,0)\approx\frac{A}{\beta_1^2-\sigma_1^2}\frac{B}{C}\exp(j\omega t) \tag{3.34}$$

$$A=\frac{\eta\,|t_1|^2\beta_1 I_0}{4k_1[1-|r_1|^2|r_2|^2e^{-2\beta_1 L_1}-2r_1r_2e^{-\beta_1 L_1}\cos(2\delta L_1)]}$$

$$\begin{aligned}B = {} & [((r-1)-(r+1)|r_2|^2e^{-2\beta_1 L_1})(b+1)e^{\sigma_1 L_1}\\ & -((r+1)-(r-1)|r_2|^2e^{-2\beta_1 L_1})(b-1)e^{-\sigma_1 L_1}\\ & +2((b-r)+(b+r)|r_2|^2e^{-\beta_1 L_1})]\end{aligned}$$

$$C=(1+b)(1+g)e^{\sigma_1 L_1}-(b-1)(g-1)e^{-\sigma_1 L_1}$$

and $r = \beta_1/\sigma_1$. This equation correctly reduces to equation (3.19) when $|r_2| = 0$, or when the surface layer is optically opaque, $\exp(-\beta L) \approx 0$. Fig. 3.12 compares the surface temperature equations for non-reflecting and reflecting layer–substrate interfaces. The surface temperature dependence on modulation frequency for

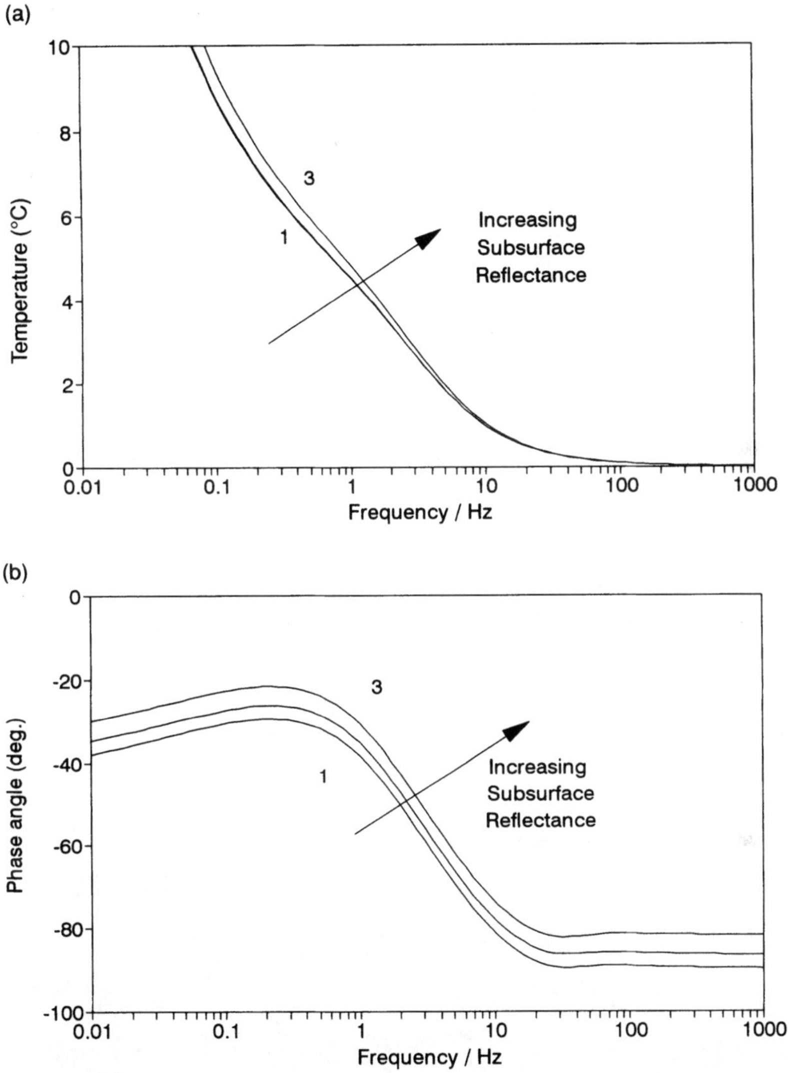

Figure 3.12 *The surface temperature (a) and phase (b) frequency dependence for the zirconia on steel coating substrate system with and without correction for substrate light reflection. Thermal properties are those given in Fig. 3.9. The optical reflectivity of the interface is set at: (1) 0, (2) 10% and (3) 50%.*

an optically reflecting substrate shows it to be reduced in magnitude with respect to the non-reflecting case. The overall temperature variation is unchanged. A similar variation is also obtained in the phase angle dependence. The phase curve of the reflecting interface with respect to the case of the non-reflecting interface depends on the complex refractive index mismatch. Three cases are presented in Fig. 3.12(b) where phase increases and decreases can be observed.

3.6 Light scattering

The assumption that the light intensity decays exponentially in a material may be incorrect for powdered or polycrystalline materials. In these types of material light scattering occurs because of the presence of the internal boundaries of the grains or individual particles, as shown schematically in Fig. 3.13. This has the effect of diffusing the incoming light beam and reducing the effective penetration of the photon flux into the sample. The scattered light energy will eventually be lost via absorption, leading to an apparent increase in the optical absorption coefficient for the medium.

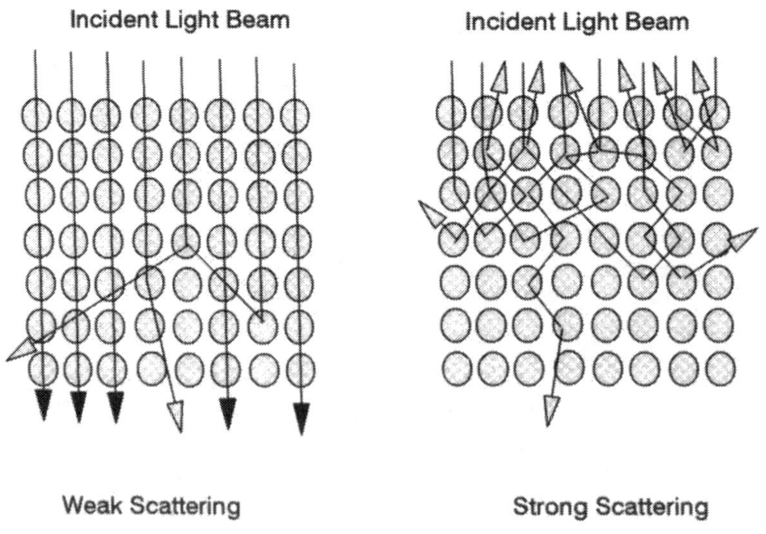

Figure 3.13 *A schematic illustration of weak and strong light scattering in a particulate solid.*

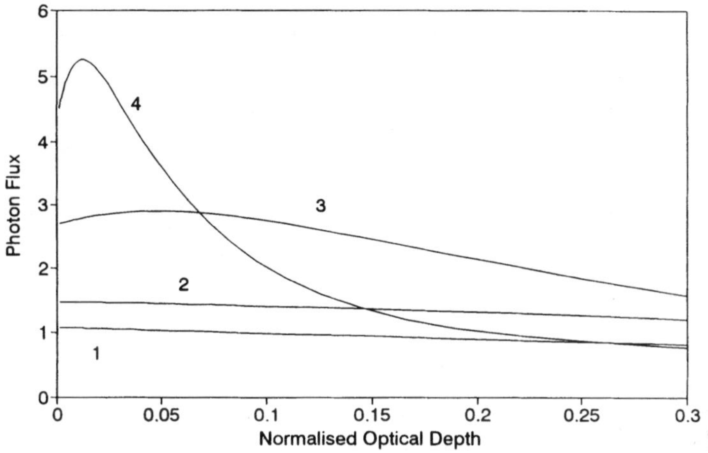

Figure 3.14 *A plot of the photon flux distribution in a sample with optical depth: (1)* $\beta_s = 0.1\beta_a$, *(2)* $\beta_s = \beta_a$, *(3)* $\beta_s = 10\beta_a$ *and (4)* $\beta_s = 100\beta_a$, $\Delta = 1.67$.

Helander (1983) derived an expression for the photon flux distribution caused by light scattering $\rho(z)$:

$$\rho(z) = \frac{I_0}{chv}\left[e^{-\beta_t z} + \frac{3}{1-m}\left(\frac{1+\Delta}{1+\Delta m} e^{-m\beta_t z} - e^{-\beta_t z} \right) \right] \tag{3.35}$$

where $\beta_t = \beta_a + \beta_s$; the subscripts a and s refer to absorption and scattering, $m = \sqrt{(3\beta_a/\beta_s)}$ and $\Delta = 2(1 + 2r_0)/3$ is a parameter determined by the boundary condition of the photon flux distribution. r_0 is the reflection coefficient of light passing from the sample interior to the outer surface. Fig. 3.14 shows the variation with normalized depth of this photon flux for several different values of β_s, the scattering coefficient. With increased scattering the photon flux deviates from an exponential decay and becomes peaked just below the sample surface. Very close to the sample surface the loss of the incident light flux with high back-scattering out of the sample can also be seen in this figure.

Using this photon flux distribution, the surface temperature of the sample assuming it to be optically and thermally thick, was found to be given by:

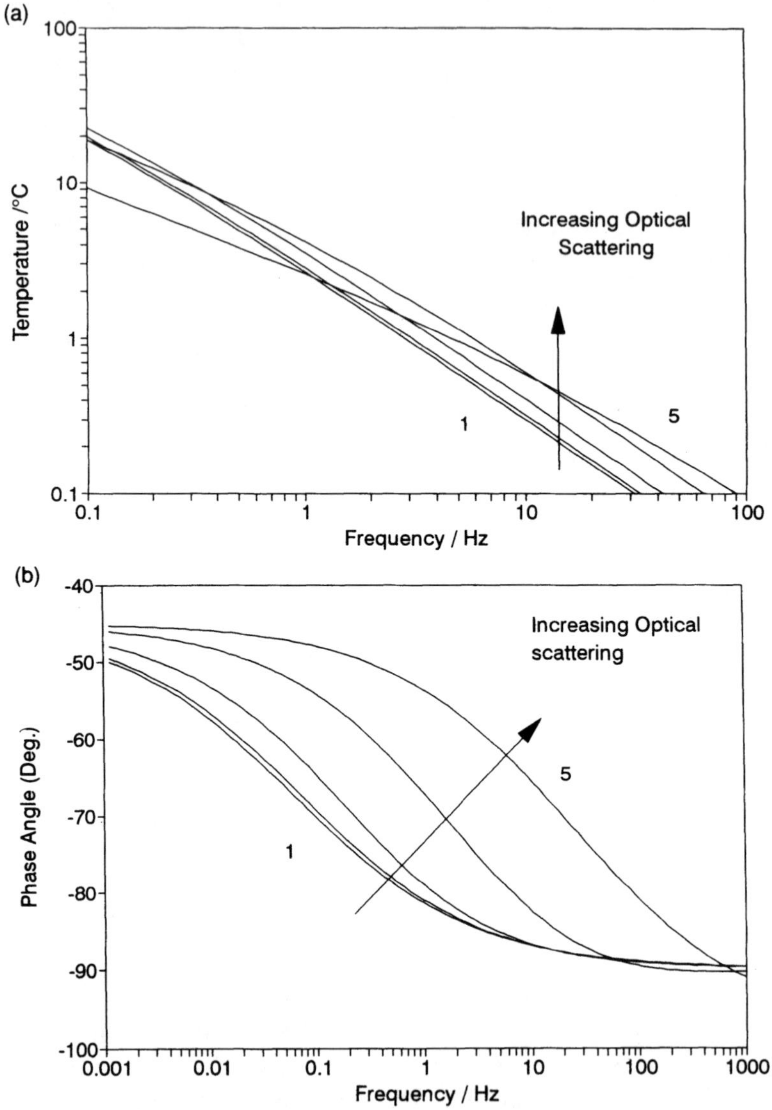

Figure 3.15 *A plot of the surface temperature (a) and phase (b) variation with frequency for a semi-infinite sample having the thermal properties of zirconia (see Figure 3.9) with scattering coefficients as indicated in Figure 3.14* β_a = 1000 m^{-1}.

$$T(\omega,0) =$$

$$I_0\beta_t\left[\frac{3(1+\Delta)}{(1-m^2)(1+\Delta m)(\sigma+\beta_t m)}+\frac{1}{(\sigma+\beta_t m)}\left(1-\frac{3}{1-m^3}\right)\right] \quad (3.36)$$

This expression for the complex surface temperature can be simplified further depending on whether the sample is weakly scattering T_{ws} or highly scattering T_{ss}:

$$T_{ws} = \frac{I_0(\beta_a+\beta_s)}{2k(\beta_a+\beta_s+\sigma)} \quad (3.37)$$

$$T_{ss} = \frac{I_0\beta_a}{2k}\left[\frac{3(1+\Delta)}{\beta_t((3\beta_t/\beta_t)^{1/2}+\sigma)}-\frac{2}{(\beta_a+\beta_s+\sigma)}\right] \quad (3.38)$$

For the weakly scattering case with β_s approaching zero T_{WS} reduces to the equation of a non-scattering sample. In Fig. 3.15 the variation of the surface temperature amplitude and phase with frequency is plotted for several scattering coefficients. For a low optical scattering coefficient the complex temperature variation follows that of the non-scattering loci (curve 1). With increasing optical scattering the temperature increases and the phase variation tends to the optically opaque limit.

There is further discussion of optical phenomena in Chapter 6 where the interpretation of optical spectra obtained photothermally is examined in detail.

References

Helander, P., 'Theoretical aspects of photoacoustic spectroscopy with light scattering', *J. Appl. Phys.*, **54** (1983) 3410–14.

Further reading

Bennett, C.A. and Patty, R.R., 'Thermal wave interferometry: a potential application of the photoacoustic effect', *Appl. Opt.*, **21** (1982) 49–54.

Mandelis, A., Sui, E. and Ho, S., 'Photoacoustic spectroscopy of thin SiO_2 films grown on (100) crystalline Si substrates', *Appl. Phys.*, **A33** (1984) 153–9.

Rosencwaig, A. and Gersho, A., 'Theory of the photoacoustic effect in solids', *J. Appl. Phys.*, **47** (1976) 64–9.

Yamasaki, S., 'Optical absorption edge of hydrogenated amorphous silicon studied by photoacoustic spectroscopy', *Phil. Mag.*, **B56** (1987) 79–97.

Yasa, Z.A, Jackson, W.B. and Amer, N.M., 'Photothermal spectroscopy of scattering media', *Appl. Opt.*, **21** (1982) 21–31.

4
Instrumentation and detection techniques

The main components of a photothermal system are:

- an excitation source;
- a modulator;
- a detector;
- a signal processing and display system.

No one combination of components is suitable for all applications. Although the nature of an application may to a large extent determine the type of source that can be used, the choice of detector may not be so clear. This is in part because the choice of photothermal system components is widest amongst the many detection schemes that have been devised to monitor photothermal and thermal wave phenomena. The characteristics of each of these detection schemes are described in this chapter, drawing attention to strengths and weaknesses for particular applications.

All photothermal systems employ a modulated source of electromagnetic radiation, usually a light source, to generate modulated heating in a sample medium. The systems rely on the absorption by the medium of electromagnetic energy and its subsequent conversion into thermal energy. This heating results in a number of physical changes in and around the sample. Figure 4.1 is a schematic illustration of the phenomena resulting from the exposure of a sample surface to a localized periodically modulated light source. In addition to a change in the temperature of the sample, the following secondary effects are also produced:

1. a modulated infrared emission from the surface;
2. a modulated thermal expansion resulting in a distortion of the surface;

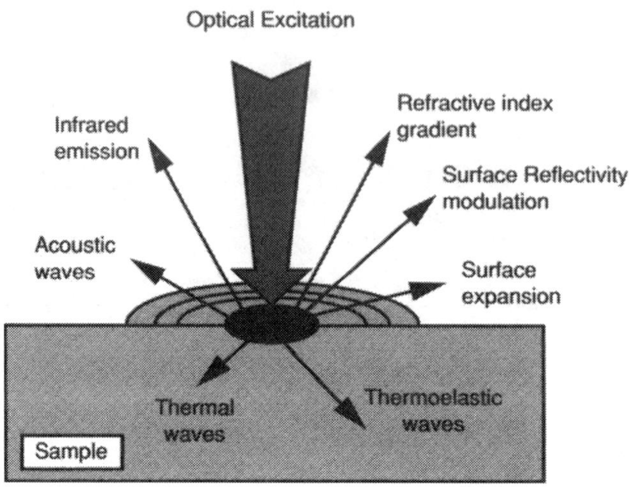

Figure 4.1 *Photothermal phenomena caused by illumination of a surface by a modulated beam of light.*

3. the generation and propagation of an acoustic wave;
4. a modulation of the optical properties of the surface such as the reflectivity;
5. a modulated refractive index gradient in any gas or other transparent medium in contact with the heated surface.

All of these effects have been used to probe the photothermal response of a medium.

4.1 Sources

Photothermal sources fall into two main classes:

1. Broad spectral range sources for spectroscopic applications;
2. Lasers.

A selection of sources and their principal characteristics is shown in Table 4.1.

The first three sources in the table are commonly used for spectroscopic applications. The high pressure xenon arc discharge lamp operates at a pressure of 50–70 atm and is a very efficient source of intense radiation from 230–2000 nm. The emission spectrum is

Table 4.1 Optical sources used in photothermal studies

Optical source	Spectral range (μm)	Power	Comment
Incoherent sources			
Xenon-lamp	0.2–2.0	0.2 W mm^{-2} sr^{-1}	High pressure arc lamp
Globar	2–30		Electric heated filament
Nernst	2–30		Electric heated

Type	Wavelength	Power/Energy	Comment
Coherent sources – lasers			
Helium–Neon (gas)	623.8 nm	0.1–50 mW	cw
Ruby (solid)	694.2 nm	0.03–100 J (per pulse)	pulsed
Carbon dioxide (gas)	10.6 μm	3–100 W	cw
Nd: YAG (solid)	1.064 μm	0.04–600 W	cw
Nd glass (solid)	1.06 μm	0.15–100 J	pulsed
Argon ion (gas)	488, 514 nm	5 mW–20 W	cw
Dye (liquid)	400–900 nm	2–800 mW	cw/pulsed (pumped by argon ion laser)
Argon fluoride (eximer)	193 nm	up to 10 W	pulsed
Hydrogen fluoride (chemical)	2.6–3.0 μM	0.01–150 W cw or 2–600 mJ per pulse	cw/pulsed
Gallium arsenide (semiconductor diode)	780–900 nm	1–40 mW cw or average per pulse	cw/pulsed
Semiconductor diode array	780–900 nm	250 mW–10W	cw

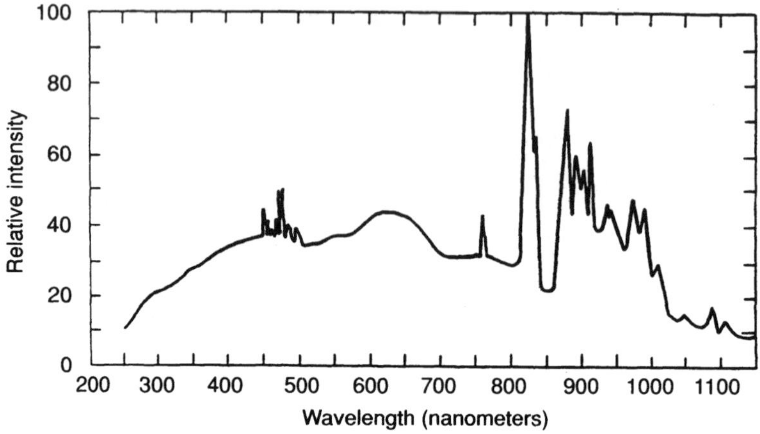

Figure 4.2 *Emission spectrum of a high pressure xenon lamp.*

primarily a continuum, with only a few intense spectral lines between 800 and 1000 nm. Figure 4.2 shows a typical emission spectrum from a high pressure xenon lamp. Power output can exceed 200 mW over all of the spectral range.

For infrared studies two commonly used sources are the globar and the Nernst glower. The globar source is an electrically heated rod of silicon carbide, operated at temperatures of 1200 K, which provides useful infrared spectral emission up to 30 μm. The Nernst glower is a more efficient source for the near to mid infrared regions, and consists of a cylindrical tube of refractory oxide that is a mixture of zirconia, yittria and thoria. As the oxides are non-conducting, they must be heated by an auxiliary source before it can be operated. Nernst glowers operate at around 1700 K and are thus more intense than globars down to 14 μm. Beyond 14 μm the globar is more intense.

Lasers are widely used because they provide the well-defined localized region of heating required for thermal wave probing techniques. In many thermal wave applications the spectral characteristics of the source are unimportant as it is simply being used as a convenient means of heating. The choice of laser is then determined by the power necessary to produce adequate heating. This is a complex function of the optothermal properties of the medium to be examined and of the sensitivity of the detection system to be employed. In most cases, however, He–Ne lasers are not powerful enough and the most widely used lasers have been argon ion lasers for cw modulated applications and Nd–YAG lasers for pulsed applications. Recently, high power semiconductor diode array lasers have become available which have the potential to displace the large cumbersome gas lasers. Their small size and ease of use have led to the development of compact portable instruments which have increased the attractiveness of photothermal and thermal wave techniques.

4.1.1 Beam focusing effects

The source (or pump) laser beam is focused in many photothermal and thermal wave systems to provide high spatial resolution and/or optical power density. In such focused systems thermal waves are generated from a point-like source, rather than an extended source, for which much of the one-dimensional heat flow analysis, introduced in the earlier chapters, is invalid. In addition, focusing can produce very high optical power densities that can raise a sample temperature high enough to cause damage.

To illustrate the effects of beam focusing we will consider samples that are optically opaque and optically and thermally homogeneous which are illuminated by a modulated laser beam focused to have a Gaussian profile with a 1/e radius of r_B. The effects of this form of illumination are to produce both a static sample temperature rise T_{DC} and a modulated component T_{AC}. In the previous chapters the static part was ignored. It will be considered here, however, because of the possibility of it being large enough to cause sample damage. For a semi-infinite solid sample:

$$T_{DC} = \frac{[1 - R(\lambda)]I_0}{2\sqrt{\pi} k r_B} \tag{4.1}$$

This static temperature rise is seen to be inversely proportional to the product of beam radius and the sample thermal conductivity. This relationship is illustrated in Figure 4.3 which shows the temperature rise calculated for a range of materials absorbing one milliwatt of laser light focused to an $r_B = 2\ \mu m$ spot. The figure shows that very high temperature rises can be produced in low conductivity materials by highly focused laser beams.

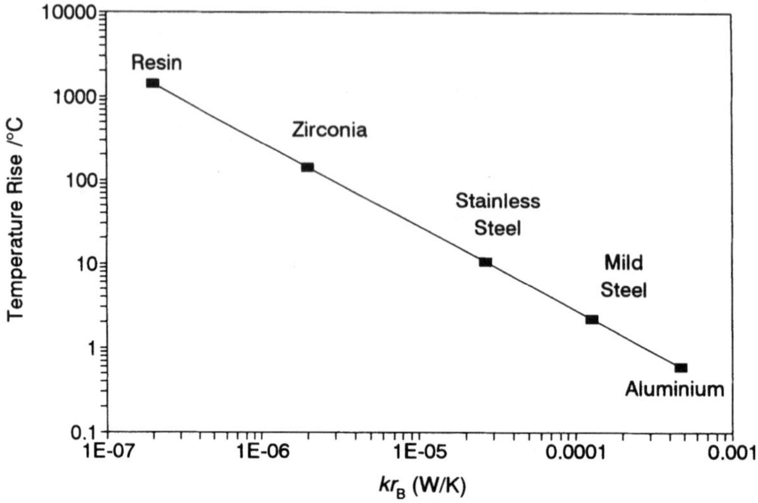

Figure 4.3 *Static temperature rise produced by the absorption of 1 mW of laser light focused to a 2 μm radius spot in the materials indicated.*

The modulated temperature change at distance r from the beam center is given by:

$$T_{AC}(\omega,r) = \frac{[1-R(\lambda)]I_0}{2\pi k}\int_0^\infty \frac{X\exp(-X^2 r_B{}^2/4)J_0(Xr)\,dX}{(X^2 + j\omega/\alpha)^{1/2}} \tag{4.2}$$

in which X is an integration variable and $J_0(...)$ the zero order Bessel function. Figs 4.4(a) and (b) show the modulation frequency dependences of the amplitude and phase of T_{AC}, at beam centre, for a range of beam radii on a stainless steel sample. At low frequencies and for small spot sizes the amplitude is independent of frequency and the thermal phase lag approaches zero. Under these conditions the induced surface temperature modulation is almost in phase with the applied periodic heating. This contrasts with the high frequency large spot radius case where amplitude falls with frequency and the phase lag approaches 45°. Under these conditions the thermal diffusion length is much smaller than the spot size and the characteristics of one-dimensional modulated heat flow are recovered. Specifically, if the beam spot diameter is >20 × (thermal diffusion length) accurate one-dimensional results will be obtained. The phase angle difference between the three-dimensional and one-

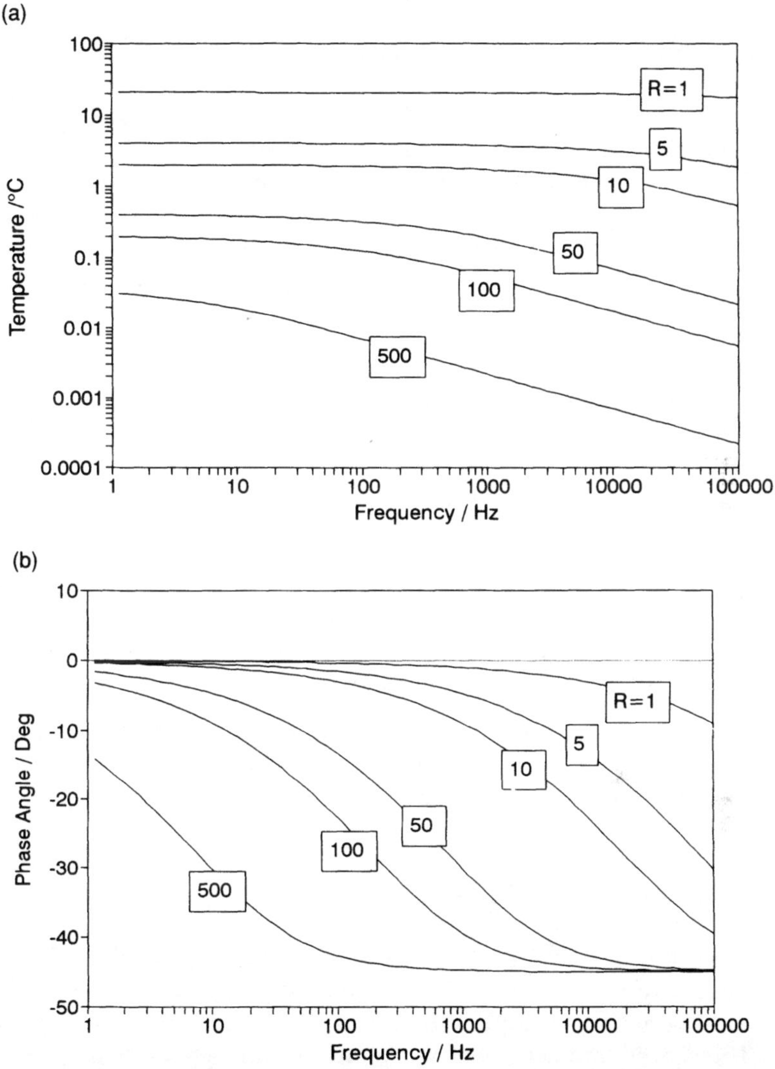

Figure 4.4 *Modulation frequency dependence of (a) the amplitude and (b) the phase of the periodic surface temperature produced in stainless steel by the absorption of 1 mW of laser light focused to a spot of radius indicated in μm.*

dimensional temperature characteristics will be less than 1°. Between the two limiting cases T_{AC} depends on frequency and spot size. It is important to note that the frequency dependent characteristics shown in Figs 4.4(a) and (b) will be superimposed on other frequency dependent phenomena that may be under investigation with a focused photothermal instrument. This can add considerably to the complexity of the analysis and to the danger of producing spurious effects caused by a change in the effective dimensionality of heat flow.

4.2 Modulation techniques

Several methods have been used to impose a temporal variation on the optical energy applied to a sample. The type of modulation can be:

1. periodic (sinusoidal or square wave);
2. transient (short, long or finite pulse);
3. frequency multiplexed (pseudo-random or frequency-modulated);
4. spatially modulated (scanned or periodic).

The first three options impose a temporal variation on the optical beam at a stationary spot on a sample. The last option produces a modulation at the sample surface by moving the heated spot.

4.2.1 Mechanical chopper

A rotating sectored blade wheel provides an inexpensive and efficient way to modulate a light beam. High quality, variable speed and low vibration noise light choppers are commercially available. They provide square wave modulation with a 100% modulation depth covering a frequency range from about 1 Hz to 20 kHz. The main limitations of the mechanical chopper are its frequency range and its restriction to square wave modulation. The high frequency limit of about 20 kHz is determined by the maximum motor speed. The low frequency limit is caused by the high phase jitter that develops at low frequencies due to the poor performance of a motor in providing a uniform rotation at very low speeds.

4.2.2 Direct electrical modulation

For some laser light sources such as CO and CO_2 gas lasers and semiconductor laser diodes, the optical output can be directly modulated by varying the electrical drive current. The gas laser systems cannot be modulated in this way at very high frequencies (>200 Hz) but semiconductor diode lasers can be modulated to very high megahertz frequencies. The depth of modulation is not in all cases the 100% achieved with a mechanical chopper. However, there are the advantages of eliminating: the need for a separate modulator, the environmental acoustic noise, and low frequency jitter problems which are present with mechanical choppers. Pure sinusoidal modulation, or almost any other desired wave form, can be produced by these direct modulation techniques.

4.2.3 Acousto-optic modulator

The acousto-optic method involves the spatial modulation of a laser beam by an acoustically formed diffraction grating, illustrated schematically in Figure 4.5. The laser beam is diffracted by the diffraction grating produced by the propagation of ultrasonic waves through a suitable crystal. The crystal material is usually dense flint glass, lead molybdate or tellurium dioxide. The power in the

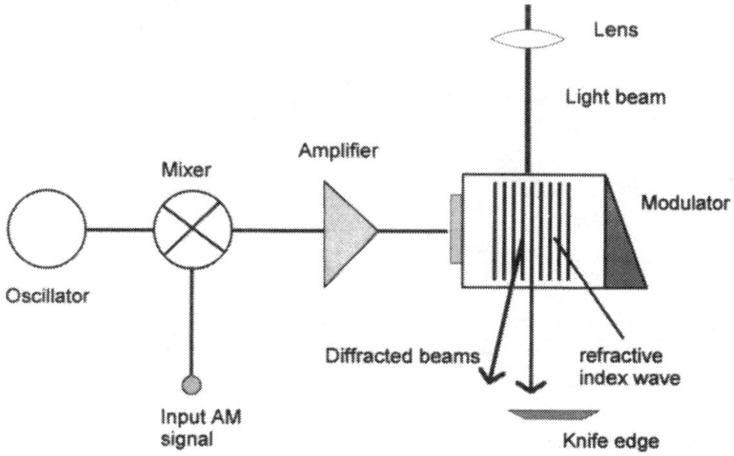

Figure 4.5 *The acousto-optic modulation process.*

diffracted output beam can be as high as 80% of the input laser power. Genuine sinusoidal intensity modulation can be obtained and acousto-optic devices can also be employed for beam scanning.

4.2.4 Electro-optic

The electro-optic method involves changing the plane of polarization of an incoming polarized laser beam in a non-linear crystal (ADP or KDP) by the application of a modulated electric field. The laser beam coming out of the crystal is passed through a polarizer which is crossed with respect to the initial plane of polarization of the laser beam as shown in Figure 4.6. The performance of an electro-optic crystal is high in the visible and near-infrared spectral region. The depth of modulation achieved can be as high as 100% and the frequency modulation range is 0–20 MHz. The laser power that can be safely used in these types of modulator is limited by the damage threshold of the electro-optic crystal.

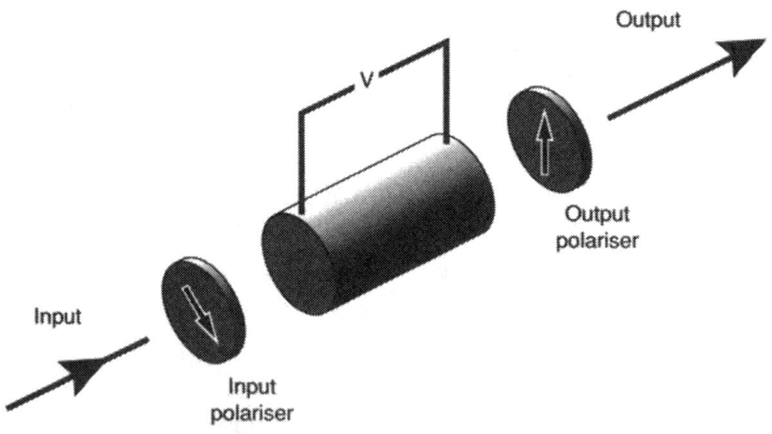

Figure 4.6 *The electro-optic modulation process.*

4.2.5 Beam deflection

The spatial modulation of a laser beam can be achieved by a variety of methods. Acousto-optic beam deflection, a rotating multiface mirror, vibrating mirror, or moving reflectors can be used to modulate or raster scan a laser beam. With the last three methods all the available energy is utilized without loss. To achieve both spatial scanning and a small optical spot size requires a careful design of the focusing optics.

Detection schemes

Three general schemes can be identified for the detection of thermal waves: acoustic, optical and thermal. Acoustic detection techniques employ either a gas condenser microphone for the detection of pressure variations in air or a piezoelectric transducer for the detection of thermoelastic waves in solid media. Optical methods for thermal wave sensing include the use of probe beams and photodetectors to monitor variations in the optical properties of a heated sample or the fluid medium adjacent to the sample. Optical methods can also be used to monitor the modulated thermal expansions of heated media by probe beam displacement techniques or via interferometry. Thermal detection methods include the use of thermocouples, thermistors, infrared detectors or pyroelectric transducers to detect thermal waves directly.

4.3 Photoacoustic (PA) detection

The photoacoustic effect has had a very significant role in the historical development of photothermal science, as was outlined in Chapter 1, and it is still a widely used method of detection. Photoacoustic signals can be generated via a variety of mechanisms. Acoustic waves can be generated by:

1. heating a gas layer adjacent to the optically heated region;
2. the periodic thermal expansion of a sample surface;
3. evaporation of volatile material;
4. the flexing vibration of a thin film sample;
5. gas evolution.

The first of these is usually the main component of the detected acoustic signal in photoacoustic studies. Contributions from the case 2 arise from porous or powder samples. The harmonic heating of a thin layer of gas close to the sample surface (known as the thermal piston model) is the primary source of the acoustic signal generation. This model predicts the gas pressure fluctuations, assuming an adiabatic gas, to be:

$$\mathrm{d}P(t) = \gamma \frac{P_o 2\pi\mu_g}{L_g T_o} \langle \mathrm{d}T(t) \rangle = \gamma \frac{P_o 2\pi\mu_g}{L_g T_o} \left(\frac{1}{2\pi\mu_g} \int_0^{2\pi\mu_g} T_s e^{-\sigma_g x + j\omega t} \mathrm{d}x \right)$$

where P_o and T_o are the ambient pressure and temperature in the cell, Figure 4.7, $\gamma = c_p/c_v$ (the ratio of specific heats at constant pressure and at constant volume) and μ_g is the gas thermal diffusion length. $\langle dT(t) \langle$ is the spatial average of the periodic temperature rise in the photoacoustic cell evaluated over a distance of one thermal wavelength in the gas, and L_g is the cell length. Evaluating this expression, the photoacoustic signal is found to be related to the sample surface temperature T_s by:

$$\mathrm{d}P(t) = K_1 K_2 e^{j(\omega t - \pi/4)}, \quad K_1 = \gamma \bar{P} / 2\sqrt{2} L_g \bar{T}, \quad K_2 = T_s(0) \quad (4.4)$$

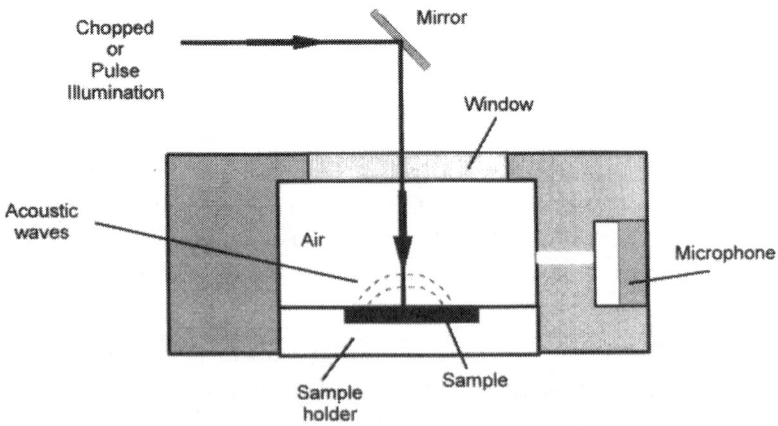

Figure 4.7 *Schematic diagram of a photoacoustic cell.*

Table 4.2 Dependence of the photoacoustic signal parameter K_2 on thermal and optical parameters

Case no.	Conditions			$K_2/I_0[I - R(\lambda)]$
1	$\mu_\beta > L_s$	$\mu_s \gg L_s$	$\mu_s > \mu_\beta$	$(1 - j)\beta L_s \mu_g \mu_b / 2k_b$
2		$\mu_s \gg L_s$	$\mu_s > \mu_\beta$	$(1 - j)\beta L_s \mu_g \mu_b / 2k_b$
3		$\mu_s \ll L_s$	$\mu_s \ll \mu_\beta$	$- j\beta \mu_s^2 \mu_b / 2k_s$
4	$\mu_\beta \ll L_s$	$\mu_s \gg L_s$	$\mu_s \gg \mu_\beta$	$(1 - j)\mu_g \mu_b / 2k_b$
5		$\mu_\beta < \mu_s < L_s$		$(1 - j)\mu_g \mu_b / 2k_s$
6		$\mu_s \ll L_s$	$\mu_s < \mu_\beta$	$- j\beta \mu_s^2 \mu_b / 2k_s$

The constant K_1 is determined by the cell characteristics (L_g, P and T) and the constant K_2 represents the sample surface temperature which depends on the applied optical intensity, and the physical properties of the coupling gas, sample and backing media.

For an absorbing layer on a non-absorbing homogeneous thermally thick substrate, the sample surface temperature is that given by equation (3.19). The six cases of the photoacoustic model, presented in Table 4.2, define the dependences of the acoustic signal on a sample's optical and thermal properties for the various relative length scales of layer thickness, thermal and optical absorption depth.

In Table 4.2 the subscripts s, g and b refer to the sample, the coupling gas and the backing material behind the sample and β is the optical absorption coefficient. Conditions 1–3 are the cases where the sample is optically translucent and 4–6 when the sample is optically opaque. Where $L_s \gg \mu_s$ the sample is thermally thick and the PA signal is determined by its thermal properties. Where $L_s \ll \mu_s$ the sample is thermally thin and the PA signal becomes dependent on the thermal properties of the backing.

The photoacoustic signal is an average measurement of the local modulated temperature rise resulting from optical heating. The design of a photoacoustic cell needs to be optimized with regard to signal detection and noise minimization. The principal noise components arise from environmental noises and secondary photoacoustic signal generation due to light absorption by the cell walls and windows. Other factors that may need consideration are: the

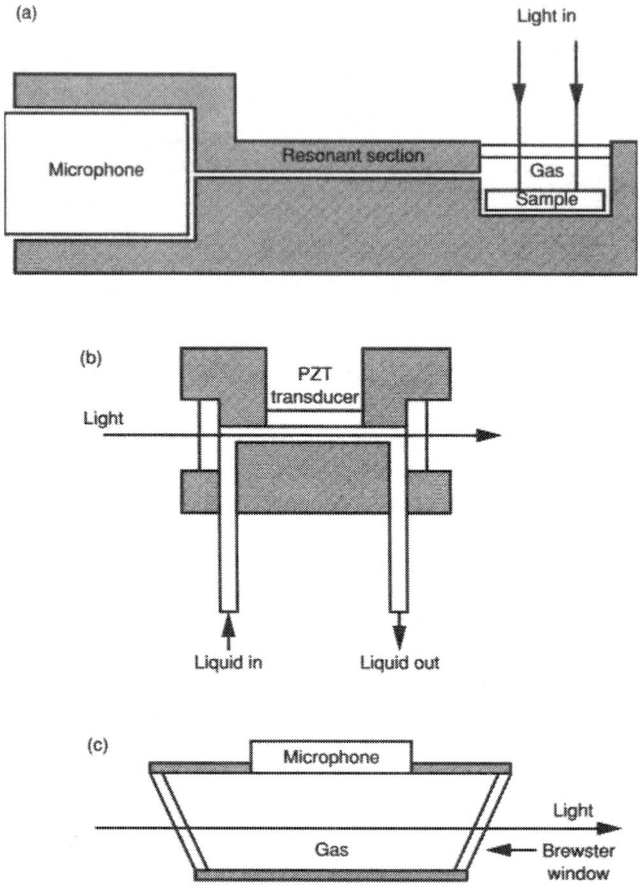

Figure 4.8 *Schematic diagrams of: (a) a resonant photoacoustic cell; or (b) a photoacoustic cell for the measurement of liquid samples and (c) a photoacoustic cell for the measurement of gaseous samples.*

acoustic response (resonances) of the cell; the type of material to be studied (solid, liquid or gas); the temperature range required and the modulation frequency range. Some further examples of photoacoustic cells are shown in Figure 4.8. Figure 4.8(a) is a cell which employs a resonant section to enhance its sensitivity at the resonant frequency. Figure 4.8(b) is a cell designed for use with liquids in which a piezoelectric transducer is in direct contact with the liquid. Figure 4.8(c) is a cell designed for use with gases which incorporates Brewster windows to reduce internal optical reflections that cause spurious signals. The photoacoustic method is particularly suitable for investigations of gases and powdered solids that have large surface area to volume ratios, as for these types of media temperature rises as small as 10^{-6} °C are detectable. A serious limitation of photoacoustic detection is the need to enclose the sample being studied in a cell. However, for spectroscopy this is usually acceptable and photoacoustic detection is the basis of photoacoustic spectroscopy (PAS) which will be discussed in some detail in Chapter 6.

4.4 Piezoelectric (PZ) detection

In this technique the photo-induced thermal expansion of a sample is sensed by a piezoelectric transducer in contact with the sample, as illustrated in Figure 4.9. The transducer converts the applied strain on its surface to an electrical voltage. Two thermal expansion terms can be identified for localized heating of a slab of material. One is due to the overall expansion of the sample caused by its average temperature rise. The other is an expansion at the

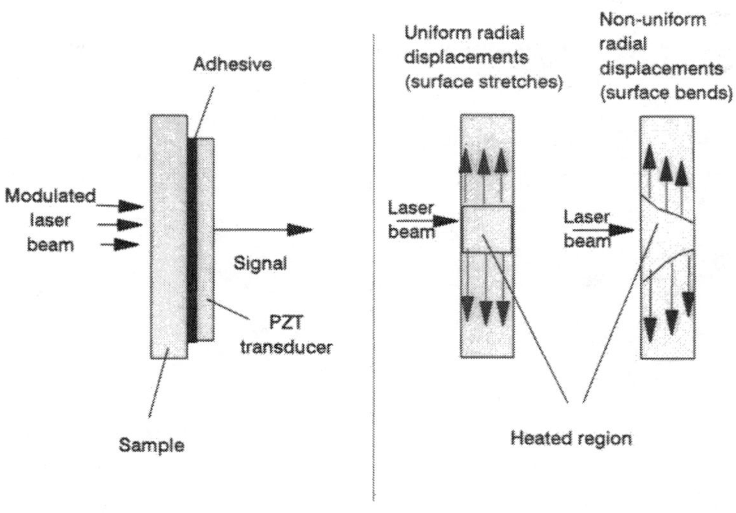

Figure 4.9 *Piezoelectric detection.*

sample surface resulting from the characteristic decay of surface temperature over a thermal diffusion length. This induces a bending of the sample whose magnitude is proportional to the average temperature gradient in the sample.

Piezoelectric photoacoustic theory provides an expression for the output voltage V from a piezoelectric transducer attached to an isotropic elastic slab of material whose thickness L_s is greater than the transducer thickness L:

$$V = M\alpha_t[\langle T(x)\rangle + (x - L_s/2)\langle \tau(x)\rangle]_{x=0,L_s} \tag{4.5}$$

where

$$M = p_z L(1+\nu)/\varepsilon_r A$$

Where p_z and ε_r are the transducer effective piezoelectric and dielectric constants along the x-direction (direction of incidence of laser beam), L and A are the transducer thickness and area, α_t and ν are the linear thermal expansion coefficient and the Poisson ratio of the sample respectively. $\langle T(x)\rangle$ is the average sample temperature and $\langle \tau \rangle$ is the average thermal gradient across the sample. The heated side of the sample corresponds to $x = 0$ and the transducer can be attached either to this side or the rear face at $x = L_s$. Expressions for the piezoelectric signal in various types of sample are listed in Table 4.3.

Where the sample is optically opaque (the photothermal heating may be taken to occur in the surface layer), the piezoelectric signal

Table 4.3 Piezo-photoacoustic signals

Condition	Signal
Thermally thick, optically opaque	$V \approx -\dfrac{2MP\alpha_t}{j\omega\rho_s c_s L_s}$
Thermally thin, optically opaque	$V \approx \dfrac{2MP\alpha_t}{j\omega\rho_s c_s L_s}$
Thermally thick, optically thin	$V \approx \dfrac{2MP\alpha_t}{j\omega\rho_s c_s L_s}(\beta L_s - 5)$

is inversely proportional to the thermal capacity of the sample, the modulation frequency and the sample thickness. When the sample is thermally thick and optically thin the piezoelectric signal also depends on the sample optical absorption coefficient, β. In general the term:

$$M\alpha_t(1/\rho_s c_s)(P/\omega) \tag{4.6}$$

controls the overall magnitude of the piezoelectric signal. P/ω is the energy deposited per cycle, $1/\rho_s c_s$ converts this energy into temperature, α_t transforms the temperature rise to a strain and M is a factor converting the strain into a voltage. The piezoelectric material used in most photothermal work is lead zirconate titanate, PZT, which provides a large M value and the capacity for high temperature operation.

The requirement of mechanical coupling between the sample and the transducer restricts the application of this technique. The piezoelectric technique does, however, overcome some of the limitations of the photoacoustic detection. It enables measurements to be made at high frequencies (>1 MHz) and it eliminates the need for a closed gas-tight cell and associated problems of acoustic cell resonances and environmental noise.

4.5 Optical beam deflection (OBD) – transverse and colinear mirage effect detection

The spatial variation of a heated material's refractive index, or that of a gas or fluid in contact with it, will deflect a laser beam. This deflection can be monitored and used as a means of detecting thermal waves generated in a sample. The change in the refractive index resulting from heating can be written as:

$$\frac{\Delta n(r,t)}{n_0(r,t)} = \frac{1}{n_0(r,t)} \frac{\mathrm{d}n(r,t)}{\mathrm{d}T} \Delta T(r,t) \tag{4.7}$$

The deflection D of an infinitesimally narrow probe beam passing through a refractive index gradient in a medium is given by:

$$D = \frac{1}{n} \frac{\mathrm{d}n}{\mathrm{d}T} \int \mathrm{d}s \nabla T(r,t) \tag{4.8}$$

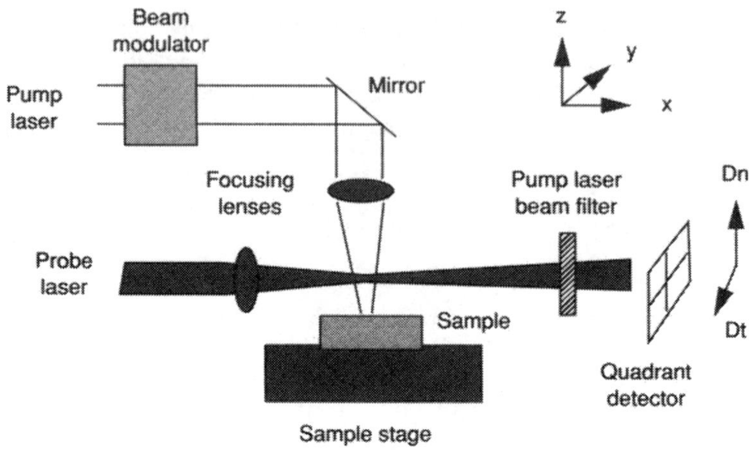

Figure 4.10 *Schematic diagram of the optical beam deflection (mirage effect) photothermal detection technique.*

For the refractive index gradient set up at a sample surface by a modulated Gaussian profile heating (pump) laser beam, using the transverse probing of the mirage deflection signal (see Figure 4.10):

$$D = -\frac{1}{n}\frac{dn}{dT}\int_{-\infty}^{\infty} \nabla T(r,t) \times d\bar{l} \tag{4.9}$$

from which two deflection components follow:

a normal component D_n:

$$D_n = \frac{1}{n}\frac{dn}{dT}\int_0^{\infty} \sigma_f(\lambda)\cos(\lambda y)E(\lambda)\exp(-\sigma_f(\lambda)z)\,d\lambda \tag{4.10}$$

and transverse component D_t, beam deflection:

$$D_t = -\frac{1}{n}\frac{dn}{dT}\int_0^{\infty} \lambda\sin(\lambda y)E(\lambda)\exp(-\sigma_f(\lambda)z)\,d\lambda \tag{4.11}$$

where σ_f is the thermal wave number for the heated, fluid or gaseous, medium. The transverse component is seen to be zero at the center of the laser heating beam where the temperature gradient is zero. This component also vanishes when the heating is uni-dimensional. The normal deflection component is sensitive to heat flow normal to the sample surface and the transverse deflection signal is sensitive to the lateral heat flow. The coefficient $E(\lambda)$ is a product of two factors: a Hankel transform of the Gaussian profile heating laser beam and an expression for the three-dimensional surface temperature. The latter is obtained by replacing the one-dimensional thermal wave vector with a three-dimensional equivalent:

$$\sigma_{1D} \Rightarrow \sigma_{3D} = \left(\lambda^2 + \sigma_{1D}^2\right)^{1/2} \tag{4.12}$$

Under the plane optical illumination of an optically opaque semi-infinite sample the probe deflection signal reduces to a single component, the normal deflection:

$$D_n = (1 - R(\lambda))I_0 \frac{L}{n}\frac{dn}{dT}\frac{\sigma_f \exp(-\sigma_f z)}{2k\sigma_s} \tag{4.13}$$

where L is the interaction length of the probe beam in the refractive index gradient. From this equation it can be seen that for a fixed interaction length, the three parameters which strongly influence the magnitude of the deflection signal are the refractive index coefficient, the fluid thermal diffusivity, which determines σ_f, and the vertical probe beam offset z. Values of these parameters are given in Table 4.4 to show that a considerable increase in signal can be effected if the deflecting medium is changed from air to a

Table 4.4 Normal probe beam deflection magnitude for uni-dimensional optical heating

Material	Thermal diffusivity ($\times 10^{-6}$ m^2 s^{-1})	Thermal diffusion (μm)	Refractive index coeff. length ($\times 10^{-4}$ K^{-1})	D_n ($\times 10^{-4}$ rad^{-1})
Air	20	800	0.01	0.1
SiO_2	0.4	110	0.1	3
H_2O	0.15	70	1	30
CCl_4	0.07	47	5	100

For 1 W of laser power absorbed at 10 Hz, L = 1 cm.

liquid. CCl_4 offers the highest gain in signal amplitude by a factor of 1000 with respect to air. In practice this magnitude of signal enhancement is not easily achieved due to probe beam alignment problems (discussed below). However, a temperature sensitivity of 10^{-7} °C is possible for mirage detection and it is clear that the use of a liquid medium should be considered in any OBD measurement.

The detection of the probe beam deflection can be accomplished by a number of methods (see Figure 4.11):

1. a knife-edge and photodetector;
2. an iris and photodetector;
3. a split photodetector (bi-cell and quadrant);
4. a lateral array position sensor.

Method 3 has the important advantages over the other methods that the measurement of deflection is independent of the position of the photodetector with respect to the sample and that laser intensity noise may be minimized by normalization. The main noise components in the probe beam deflection set-up arise from: vibrations; probe beam intensity noise and pointing instability, and electronic noise. In a well optimized setup the beam pointing and shot noise are the limiting noise sources. The technique is limited to applications on smooth planar surfaces. Curved surfaces or

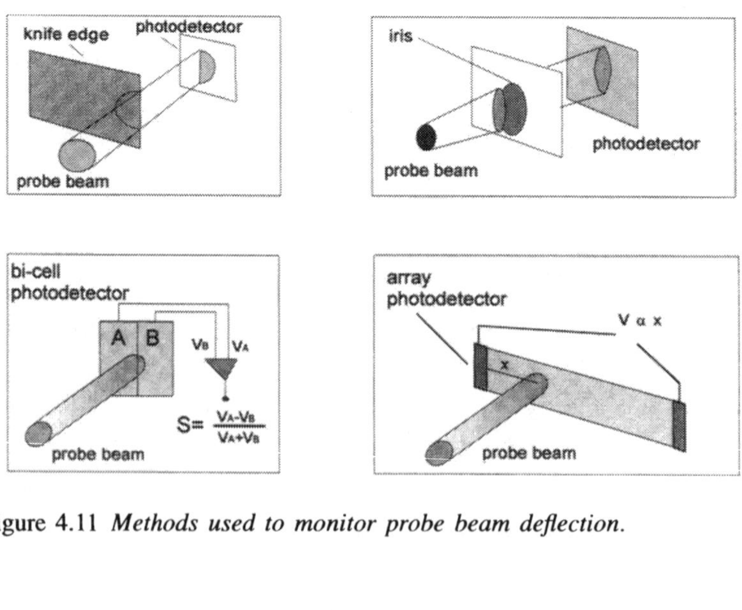

Figure 4.11 *Methods used to monitor probe beam deflection.*

surface roughness significantly restrict the successful application of this detection technique. If the sample is transparent, the probe beam can pass through the refractive index gradient produced in the sample. The exploitation of this mode of operation for the study of semiconductors and gases is covered in Chapters 8 and 10.

4.6 Modulated optical reflectance (MOR)

Temperature modulation can also lead to a modulation of a sample's optical reflectivity via the modulation of its complex refractive index N (see Chapter 3). The change in reflectance can be related to the sample surface temperature as follows:

$$\frac{\Delta R}{R_0} = \frac{1}{R_0}\frac{\mathrm{d}R}{\mathrm{d}T}\Delta T \tag{4.14}$$

where $(1/R_0)\mathrm{d}R/\mathrm{d}T$ is the coefficient of thermal reflectance of the sample. For most solids this coefficient ranges from 10^{-4} to 10^{-6} per degree.

A typical modulated optical reflectance (MOR) detection scheme is shown in Figure 4.12. The pump beam which can be either periodically or pulse modulated is expanded and focused with a microscope objective onto the sample surface. A probe beam is also allowed to illuminate the sample colinearly after passing through a polarizing beam splitter and a quarter-wave plate. The retroreflected probe beam passes through the quarter-wave plate again and now, since its polarization is rotated by 90° with respect to the incoming beam, it is reflected by the polarizing beam splitter towards the photodetector. The latter is suitably shielded from any modulated pump laser light and under filled by the probe beam so that the detector only monitors the reflectance signal and is insensitive to the small variations of probe beam size or reflected angle that can also occur from thermal lens and thermoelastic deformation effects at the sample surface. A suitably optimized system is capable of detecting periodic modulated reflectance signals of 10^{-7} $\mathrm{Hz}^{1/2}$. Commercial systems are employed industrially for semiconductor wafer evaluation, see Chapter 8.

Some practical problems in the usage of this technique include the need for very accurate alignment of the probe and pump beams, efficient screening of the reflected probe beam from re-entering into

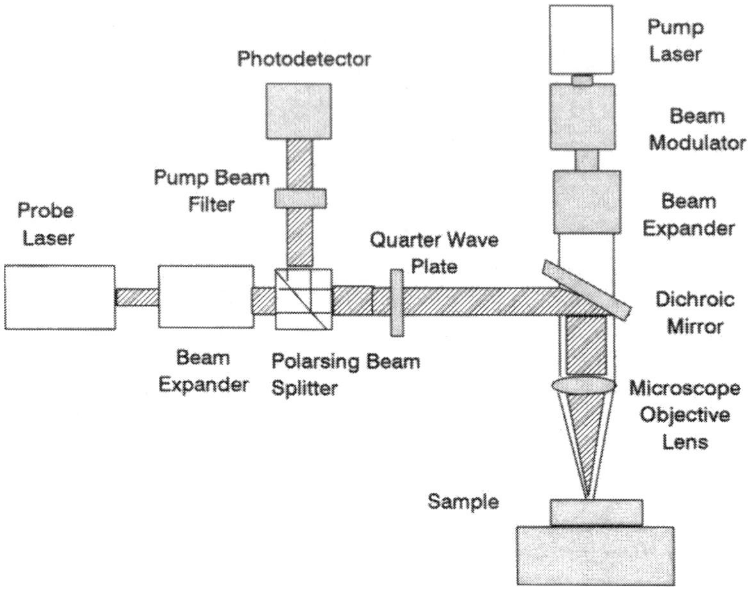

Figure 4.12 *Schematic diagram of a modulated optical reflectance (photo-reflectance) system.*

its laser cavity, and loss of reflected probe beam signal from rough or unpolished sample surfaces. However, to obtain good MOR signals it is not necessary that the sample is optically flat. An important requirement is that the reflection from the spot illuminated by the probe beam is high, as heating can be generated from a polished or a rough sample surface. So, providing a highly reflective spot can be found on a sample, MOR measurements can still be made by scanning the heating beam only while the probe beam location is fixed. Another solution to the rough surface problem would be to monitor the scattered probe laser illumination.

4.7 Photothermal displacement (PTD)

Modulated heating of a sample results in a synchronous sample expansion. The optical detection of the resulting surface displacement by an interferometer or a bi/quad-cell beam position sensor provides a highly sensitive, localized and non-contact method of

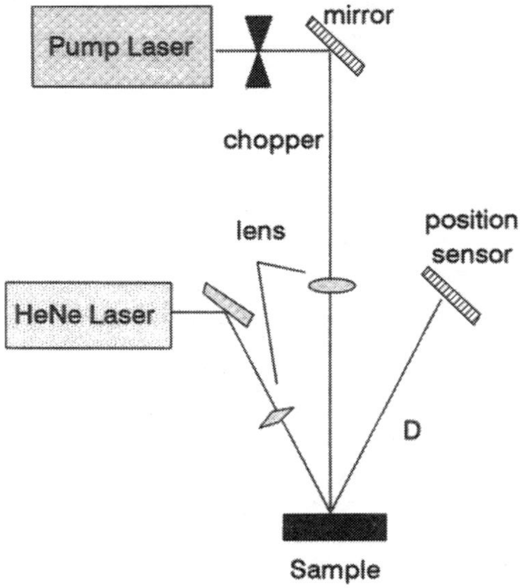

Figure 4.13 *Schematic diagram of photothermal displacement detection system employing a bi/quad-cell beam position sensor.*

sample characterization. Figs 4.13 and 4.14 illustrate these two displacement detection schemes. In the interferometric scheme the sample serves as one arm of a conventional Michelson interferometer. The mirror used in the other arm is mounted on a piezoelectric transducer for signal stabilization. The pump beam modulates the surface displacement at a frequency ω_1 and a small oscillation is applied to the mirror through the piezoelectric transducer at frequency ω_2. The signal at the photodiode has the form:

$$S = S_0 \cos(4d\pi / \lambda) + \text{constant} \qquad (4.15)$$

where λ is the wavelength of the probe laser and

$$2d = 2[d_0 + d_1 \cos(\omega_1 t) + d_2 \cos(\omega_2 t + \phi)] \qquad (4.16)$$

where d_0 is the static path difference (controlled by the voltage applied to the transducer), d_1 is the amplitude of photothermal displacement and d_2 is the amplitude of mirror oscillation (controlled by the ac voltage applied to the piezoelectric transducer). If the

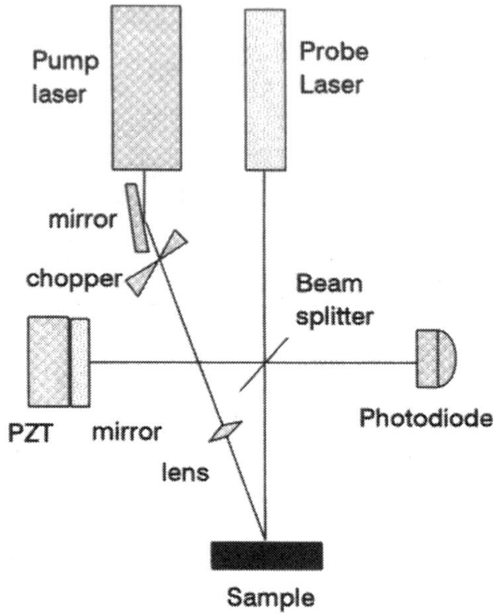

Figure 4.14 *Schematic diagram of an interferometric photothermal displacement detection system.*

amplitudes d_1 and d_2 are kept small (< 7.5 nm for 1% accuracy with a HeNe probe) then the signal at ω_1 is:

$$S(\omega_1) = S_0(2\pi d_1 / \lambda)\sin(4\pi d_0 / \lambda) \qquad (4.17)$$

The dc path difference is kept near $n\lambda/2$ by a servo loop maintaining a zero signal at $2\omega_2$ by adjusting the dc bias of the piezoelectric transducer. Active stabilization of the interferometer eliminates signal drift resulting from thermal fluctuations and acoustic disturbances in the interferometer elements.

In the beam deflection scheme the modulated displacement of a probe beam reflected from the expanding and contracting heated region depends on the local slope of the displacement. The deflection is detected by a position sensitive bi/quad cell photodiode. The deflection signal at the position sensor is given by:

$$S = \chi\left[2D\frac{\mathrm{d}u_z}{\mathrm{d}r} + 2u_z r_0 \sin\theta\right] + \text{smaller terms} \qquad (4.18)$$

where D is the distance of the sample to the position sensor, u_z surface displacement, θ is the angle of incidence of the probe beam and χ is the position sensor sensitivity. The first term is the deflection by the slope of the displacement. The second term is the small deflection of the probe beam by the vertical displacement. This term is smaller than the first by a factor D/a (typically 5000) where a is pump beam radius. To a good approximation the deflection signal is proportional to the displacement slope $\mathrm{d}u_z/\mathrm{d}r$.

The main parameters limiting the sensitivity of displacement measurements in both of these detection schemes are the probe laser intensity and pointing noise. The current sensitivity limit for an interferometer is typically 4×10^{-4} nm $\mathrm{Hz}^{1/2}$. For the beam deflection method a slope sensitivity of 1×10^{-9} $\mathrm{Hz}^{1/2}$ is expected. To maximize the displacement signal it is necessary to employ highly focused pump and probe beams. The latter should be smaller than the pump beam to allow easier data interpretation. In the beam deflection scheme sensitivity can be enhanced by maximizing the distance between the sample and the position sensor. The upper limit on this distance is determined by beam path stability and the need to maintain a probe beam spot that is smaller then the active element in the position sensor. Vibration noise sources may be minimized by using a vibration isolated optical table and optical components mounts. A particular feature of this optical detection scheme is that it can be operated in a vacuum unlike the photoacoustic and mirage photothermal techniques.

4.8 Infrared detection – photothermal radiometry (PTR)

Of all the thermal wave detection schemes, those based on the infrared detection of a sample's temperature perturbation are becoming the most widely exploite, the principal reasons being that infrared detection offers a simple means of monitoring sample temperature that is also robust, non-contacting and compatible with many industrial requirements.

Any material emits electromagnetic radiation when its temperature is above absolute zero. The spectral variation of the emitted radiation is given by Planck's radiation formula for a perfect emitter:

$$W(\lambda,T)=\frac{C_1\lambda^{-5}}{\exp(C_2/\lambda T)-1} \tag{4.19}$$

where T is absolute temperature, $C_1 = 3.74\times10^4$ W μm^4 cm^{-2} and $C_2 = 1.44\times10^4$ μm K are the first and second radiation constants respectively and λ is the wavelength (μm).

Figure 4.15 shows that a body at room temperature emits radiation covering the near and mid-infrared spectral region, peaking near 8–12 μm. The total radiation emitted is the integral of emittance over all spectral wavelengths.

The change in thermal emission produced by modulated sample heating can be derived from the Stefan–Boltzmann law (equation (2.5)):

$$\begin{aligned}\delta W &= \varepsilon\sigma_{\mathrm{SB}}\left[(T_0+T_{\mathrm{dc}}+T_{\mathrm{ac}})^4-(T_0+T_{\mathrm{dc}})^4\right]\\ &\approx 4\varepsilon\sigma_{\mathrm{SB}}(T_0+T_{\mathrm{dc}})^3T_{\mathrm{ac}}\end{aligned} \tag{4.20}$$

To a first order, the emitted radiation is directly proportional to the modulated component of the sample surface temperature T_{ac} and to the cube of its local static temperature, (T_0+T_{dc}). For a broad-band infrared detector of area A and collection angle Θ, and assuming

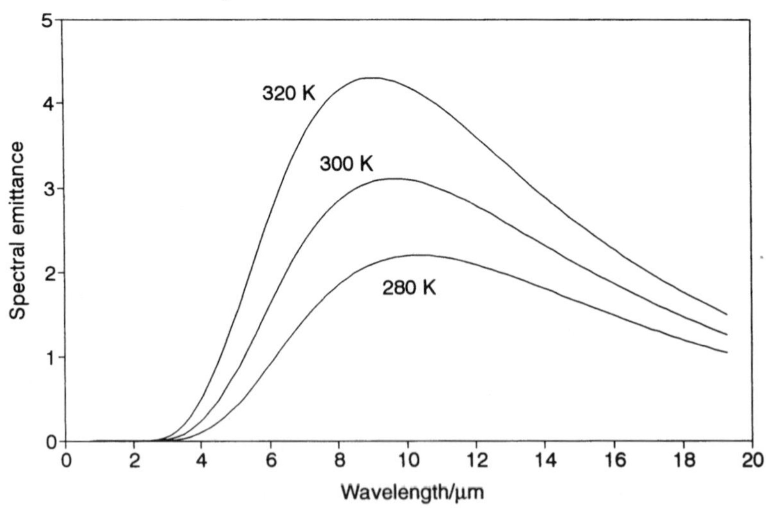

Figure 4.15 *Black body radiation spectrum at temperatures near ambient.*

Lambertian thermal emission from a sample surface the PTR signal may be written as:

$$S_{\text{ptr}} \approx 4\varepsilon\sigma_{\text{SB}}(T_0 + T_{\text{dc}})^3 A \sin^2 \Theta T_{\text{ac}} \tag{4.20a}$$

The use of broad-band thermal detectors is suitable only when strong signals are available. An example of a broad-band thermal detector is the pyroelectric infrared sensor. In this device a thin slice of a pyroelectric material is placed in thermal isolation from its surroundings. Two opposite faces of the pyroelectric material are metallized to form electrodes, making the pyroelectric detector a capacitor with the pyroelectric material acting as the dielectric. The surface facing the incoming radiation is darkened to improve optical absorption over a broad spectral range. A change in voltage across the detector electrodes is produced by the change in polarization caused by the heating due to the absorption of radiation incident on the darkened electrode. Owing to the darkening of the electrode plate, the detector is sensitive to all wavelengths of the emitted thermal radiation. In addition, because it generates a signal by the heating of the pyroelectric material, the detector requires no cooling for its operation. Commercial pyroelectric detectors are available complete with low-noise FET preamplifiers. The preamplifier may produce a high responsivity but at the expense of a limited frequency response. At high frequencies the detector signal decays inversely with increasing frequency.

Semiconductor detectors (classed as quantum detectors), cooled by liquid nitrogen or thermoelectrically, are more sensitive and operate over a wider electrical bandwidth than pyroelectric detectors. An appropriate expression for the infrared signal is then:

$$S_{\text{ptr}} = \varepsilon A \sin^2 \Theta T(\lambda_{\text{d}}) R(\lambda_{\text{d}}) \frac{\partial W}{\partial T}(\lambda_{\text{d}}) \Delta\lambda T_{\text{ac}} \tag{4.20b}$$

where $T(\ \ )$ is the spectral transmission efficiency of the collecting optics, $R(\ \ )$ is the spectral sensitivity of the detector and $\partial W / \partial T$ is the temperature derivative of Planck's radiation function.

Quantum detectors have sensitivities in a given spectral region $\Delta\lambda$ around a detection wavelength λ_{d}. Two commonly used semiconductor detector materials for general infrared detection are InSb and $CdHg_xTe_{1-x}$ The former provides a detection sensitivity in the 2–5.5 μm spectral region and the latter in the 2–12 μm spectral region, tunable by adjustment of the stoichiometric parameter x.

These devices rely on the generation of free carriers by the absorption of photons in semiconducting material with a band gap appropriate to the radiation to be detected. Hence their spectral responses are selective. These detectors require cooling to avoid the generation of a photocurrent due to the absorption of thermal radiation emitted by the detector itself at ambient temperature. The semiconductor detectors offer both a high sensitivity (indicated by the detectivity D* figure of merit) and a high frequency bandwidth (typically MHz). The bandwidth of a pyroelectric infrared sensor is only a few kilohertz.

As mentioned earlier a large number of experimental schemes have been developed for photothermal materials characterization using infrared detection. The detection schemes may be grouped with respect to the localization of infrared sensing: (a) at a point; (b) along line or (c) across an area. Systems have been developed which may employ: periodic; chirp; pulsed or step optical heating: at a point; along line; around a ring or across an area of a sample surface or its rear, with matched point, line and area infrared detection at the heated region or at some point close to it. Systems based on thermal imaging cameras are presently under rapid development. and will be discussed further in Chapter 7.

The successful use of infrared detection in a photothermal application depends on maximizing the infrared radiation collected by the detector and on minimizing the incidence of the excitation source radiation on the detector, see Figure 4.16. The first of these requirements can be satisfied by the use of suitable infrared collecting optics. Because infrared lenses are particularly expensive, mirrors are widely used. There are a wide range of commercially available ellipsoidal mirror components that are ideally suited to this application. However, the use of focusing optics does have the disadvantage of making the detector system sensitive to sample detector set-up distance. It must also be noted that although infrared detection provides a genuine non-contact sensing system, it is usually necessary to place the detector close to the sample to produce an adequate signal.

The problem of shielding the detector from the excitation source can be overcome by the use of filters and by arranging the geometry of the system to prevent the detector from being exposed to specularly reflected source radiation. These problems become particularly acute where reflective metallic surfaces are being examined. For these, very little of the source radiation is absorbed and

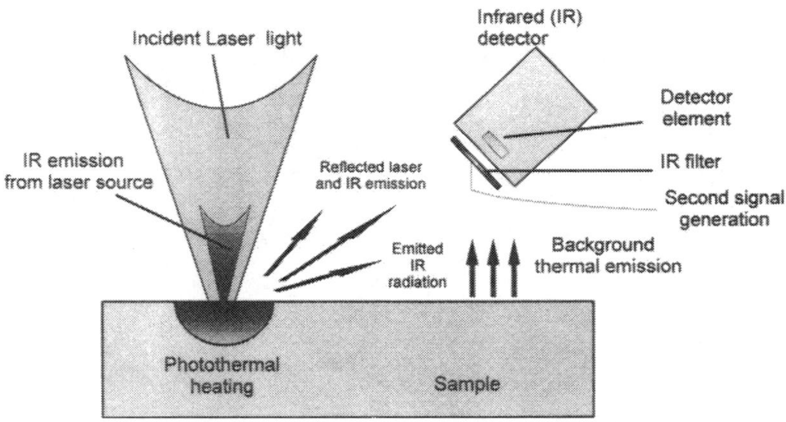

Figure 4.16 *Schematic diagram of the generation and detection of infrared radiation in a photothermal experiment.*

converted into heat and most of it is reflected or scattered, potentially towards the detector. In such applications conventional germanium filters, which remove visible source radiation by absorption, may become heated by the source to produce a spurious secondary heat source. This problem has been significantly reduced by specially designed interference filters which reflect, rather than absorb, visible light whilst being highly transmissive to infrared.

4.9 Photo-pyroelectric (PPE) detection

Pyroelectricity is the manifestation of a spontaneous temperature dependent electrical polarization. This effect is exhibited only by solids that satisfy certain crystallographic requirements, such as no center of symmetry in the crystal lattice and not more than one axis of rotational symmetry. Ferroelectric materials are generally pyroelectric. Pyroelectric materials exhibit the property that the polarization vector is a function of temperature. This property results in a temperature dependent potential difference occurring between two, properly chosen, opposite surfaces of a pyroelectric material.

In the photo-pyroelectric (PPE) photothermal technique, the modulated or pulsed temperature rise induced in a material is detected by a pyroelectric film sensor attached to the sample. The film can be attached either at the back of the sample or at its surface.

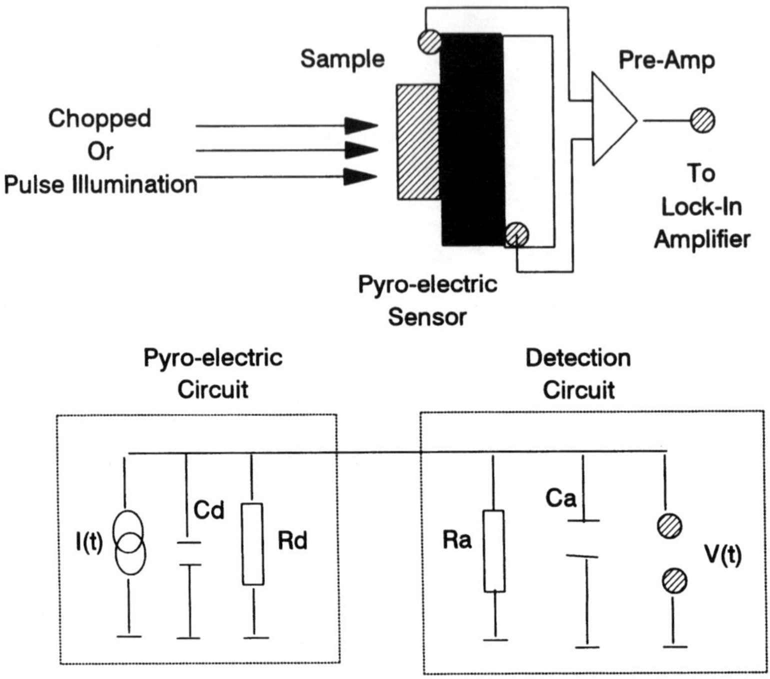

Figure 4.17 *Schematic diagram of photo-pyroelectric detection and the corresponding electrical equivalent circuit.*

A typical PPE experimental set-up is shown in Figure 4.17, together with the relevant electrical circuit model. The output voltage V due to a harmonic heating of the pyroelectric sensor is given by:

$$V = -\frac{j\omega p \tau_e L_p \langle \Delta T \rangle}{\varepsilon_p (1 + j\omega \tau_e)} \exp(j\omega t) \tag{4.21}$$

where p is the pyroelectric constant, L_p is the film thickness, τ_e is the electronic time constant:

$$\tau_e = \frac{(C_p + C_d)}{(1/R_d + 1/R_p)} \tag{4.22}$$

ε_p is the permittivity and $\langle \Delta T \rangle$ is the average temperature rise in the thin film sensor:

$$\langle \Delta T \rangle = \frac{1}{L_p} \int_0^{L_p} T_d(x,t)\,dx \tag{4.23}$$

where $T_d(x, t)$ is the temperature fluctuation in the pyroelectric film.

The determination of this temperature requires the solution of the heat diffusion equation for a system consisting of: the air layer; the sample; the coupling material between the sample and pyroelectric film; the pyroelectric film and the backing layer.

Some simple expressions which can be derived for the PPE voltage under certain limiting conditions are presented in Table 4.6. For a thermally thick and optically opaque sample layer the PPE signal is predicted to decay exponentially with sample thickness L_s. When the sample layer is translucent such that $\mu_\beta > \mu_s$ the PPE signal is linearly proportional to the sample optical absorption coefficient, β_s.

The principal advantages of the PPE technique when compared with the photoacoustic detection method include: a flat frequency response over a relatively large bandwidth; a simple sample cell and the elimination of noise problems arising from the cell walls and windows.

The choice of the pyroelectric sensor material depends on the application intended. The optical, electrical, thermal and mechanical properties need to be optimized to achieve the best overall performance. In PPE measurements a large value of the pyroelectric coefficient is required together with a low electrical conductivity or large sensor resistance to prevent leaking of the pyroelectric current. A small dielectric constant is also required to convert the induced charge into as high a voltage as possible. The thermal requirements of the sensor include a low density and specific heat capacity. The mechanical requirements include flexibility of the sensor to allow attachment to the sample. A commonly used PPE sensor for photothermal studies is the polymer electret β-polyvinyldielene fluoride (PVDF). This material is commercially available, at very low cost, in thin film sheets, with metal film electrodes deposited on each side. The material offers a high pyroelectric figure of merit (about 9) and good temperature stability up to 200°C.

Table 4.5 Expressions for the PPE signal variation, the sample system is shown in Fig. 4.17

PPE-film	Sample	Signal	Comment
Opaque and thermally thick	Opaque and thermally thick	$V(\omega) \approx A \dfrac{\eta_s \exp(-\sigma_s L_s)}{\sigma_p(k_p\sigma_p + k_s\sigma_s)}$ $A = \dfrac{pI_0}{2\varepsilon\varepsilon_r}$	Exponential signal decay with sample thickness, no optical information
Opaque and thermally thin	Thermally thick $\mu_\beta > \mu_s$	$V(\omega) \approx A \dfrac{\eta_s \beta_s \exp(-\sigma_s L_s)}{\sigma_p \sigma_s(k_p\sigma_p + k_s\sigma_s)}$	Signal proportional to optical absorption coefficient β_s
	Thermally thick and optically thin	$V(\omega) \approx A \dfrac{\eta_p L_p(1-\beta_s L_s)}{(k_b\sigma_b + k_s\sigma_s)}$	Signal proportional to $1 - \beta_s L_s$ optical transmission

4.10 Signal processing

A pervasive problem in all experimental disciplines is the extraction of good quality data from noisy signals. The method of signal recovery used in a particular photothermal/thermal wave system is dictated by the type of modulation used in generating the signal. In general the following methods can be applied to enhance the signal-to-noise (S/N) ratio: low pass filtering; ensemble averaging; phase sensitive detection; box car integration and mathematical methods (curve fitting, digital filtering).

To quantify signal quality, the S/N ratio is used. This is frequently defined as:

$$\text{S/N} = \frac{\text{signal amplitude}}{\text{standard deviation of signal}}$$

Degradation in signal quality can result from a variety of sources, such as: electromagnetic interference from power line noise; fluorescent lights; neighboring electronic instruments etc. Other contributions are intrinsic to the noise sources associated with the type of detector used or the method of detection employed.

4.10.1 Phase sensitive detection

For periodic modulation, phase-sensitive detection (lock-in amplification) can be utilized to process the detected signal. The technique is extremely efficient for extracting very weak harmonic signals contaminated with noise (–100 dB). Figure 4.18 is a schematic diagram illustrating the principle of operation of this detection scheme. Synchronous detection is accomplished by a multiplier followed by a low pass filter. The signal is multiplied with the reference, and only the low frequency components of the multiplication are filtered to the output by the low pass filter, giving a response only to signals in the vicinity of the reference frequency. All other signals and noise of the rest of the spectrum are rejected. The result is an effective narrowing of the noise bandwidth, yielding an improvement in the S/N ratio. The parameter that controls the narrowing of the bandwidth is the time constant of the low pass filter. Increasing the time constant improves the detection of low signal amplitudes (i.e. low thermal wave signals), but also degrades the system response time.

In most photothermal systems two-phase sensitive detectors (two-phase lock-in amplifiers) are used to provide a vector signal output comprising the amplitude and phase angle of the photothermal signal with respect to the reference. Another advantage of these two-phase

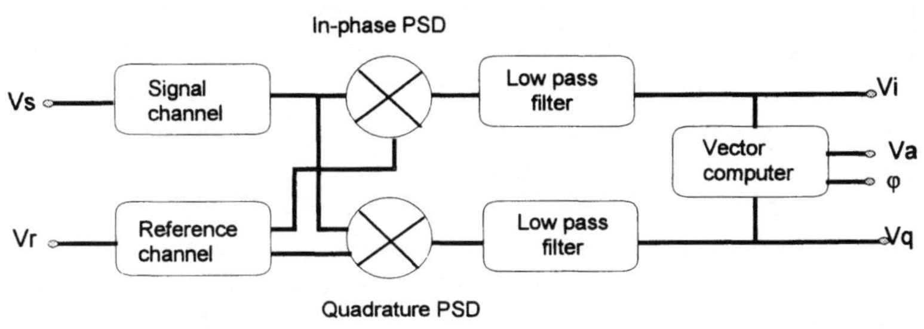

Figure 4.18 *The principle of phase-sensitive detection.*

systems is that they can be used to provide the in-phase and out-of-phase signal components. The latter is a signal component derived with the phase of the reference signal shifted by 90°. This out-of-phase component can be an important parameter to monitor as it is sensitive to signal components generated deep in a sample. It provides a means of depth profiling by reference phase angle adjustment rather than by varying the modulation frequency; an example of the use of this technique is included in Chapter 6.

A full modulation frequency spectrum of amplitude and phase is often required for quantitative applications. This may cover several frequency decades and long acquisition times may be required to obtain the necessary data points. Transient thermal excitation methods were developed to overcome these limitations of the cw photothermal technique.

4.10.2 Transient excitation

For pulsed and other transient forms of photothermal excitation the signal recovery requirements necessitate wide bandwidth detection electronics that remain linear over many orders of magnitude of signal amplitude. The implementation of digital signal acquisition with high resolution, low noise and variable time delays has provided significant improvements over traditional analog signal acquisition methods. Enhancements of the signal-to-noise ratio can be made by a variety of methods. Averaging may be performed over a number of pulses (N) to effect an increase in S/N of $N^{1/2}$. Having obtained this averaged signal a Fourier filter can be applied to remove unwanted frequency components in the detected signal. If the signal is known to exhibit a monotonic variation then a curve fitting technique can also be applied to obtain a smooth signal. These methods can also be applied over a pre-selected time interval, in an analogous fashion to the box-car technique, for further signal enhancement. Alternatively, if the transient is slow and long enough, say >1 s, the photothermal signal can be chopped and phase sensitively detected for high noise rejection.

Where transient techniques are used to obtain rapid broad spectral information about a photothermal or thermal wave phenomenon, it must be recognized that the optical excitation energy is effectively spread over a broad range of modulation frequency components and that it will not be possible to approach the S/N ratio achievable for a particular frequency component by a cw technique, in which all the excitation energy is concentrated in that frequency component.

4.10.3 FFT analysis

If the modulation applied to the excitation source is of a square wave type, for example such as produced by a mechanical chopper, a fast Fourier transform (FFT) analyzer could be used as a phase sensitive detector to simultaneously track several harmonic frequency components ω, 3ω, 5ω, 7ω . . . This provides a much greater amount of information than derived by a single lock-in amplifier during the same measurement time. The upper frequency limit is set by the available signal-to-noise ratio for the components. Typically up to five harmonic component signals can be readily measured with an FFT analyzer.

Another application of the FFT analyzer is in the recovery of photothermal signals generated by a wide-band frequency excitation using a cross-correlation signal processing method. Frequency-multiplexed photothermal techniques were developed to overcome the long acquisition times inherent in single frequency excitation techniques and the wide bandwidth and high source intensity necessary for pulsed photothermal investigations. These wide-band techniques utilize excitation with a waveform whose input autospectrum is flat over the photothermal response bandwidth. The input autospectrum $G_{xx}(f)$ measures the distribution of signal energy as a function of frequency and for deterministic signals is computed as the square of the magnitude of the Fourier transform of the input signal. Unlike the pulsed photothermal excitation method, which delivers all of its Fourier components to the sample in phase at time $t = 0$, wide-band excitation waveforms utilize random, pseudo, or quadratic phase modulation of the individual Fourier components. The resulting waveform applies excitation energy to the sample over a much greater duty cycle, and at far lower peak amplitudes. The result is a dramatic reduction in the overall peak power and the linear response requirements of the detection system.

In the study of linear systems excited by deterministic signals, the system response $Y(t)$ is recovered as the convolution of the input waveform with the impulse response. If the input signal is stochastic, the input and output signals are characterized by their correlation functions rather than their instantaneous values. The system response is then recovered as:

$$R_{XY}(\tau) = \int_0^\infty h(u) R_{XX}(\tau - u)\,du \tag{4.24}$$

If the power spectrum of the input waveform is wide-band and flat relative to the photothermal response spectrum, we have the condition $R_{XX}(\tau) = \delta(\tau)$ so that $R_{XY}(\tau) = h(\tau)$. Therefore, a measurement of the cross-correlation function gives a good approximation of the impulse response of the system. This measurement is normally carried out experimentally using an FFT analyzer or hard wired correlator.

Several classes of waveform are available with nearly flat power spectra, including wide-band random noise, pseudo-random binary sequences and linear frequency sweeps. Linear frequency sweeps have been shown to exhibit the best overall spectral flatness and frequency-domain dynamic range of the three types of waveform.

4.11 Sample scanning

Sample scanning is necessary when different regions of a sample are to be interrogated, compared or imaged. Thermal wave techniques which employ localized excitation, such as are used for thermal imaging or non-destructive inspection, require a scanning and image display system. Whilst the scanning of the excitation (pump) laser beam can be readily achieved using scanning mirrors, it is often difficult to scan the detection system simultaneously, maintaining the often critical relationship with the pump beam. For this reason the alternative of moving the sample with respect to a fixed pump probe system is often chosen. Stepping motor driven translation stages can be used to provide step movement from 1 μm to a few millimeters. These step ranges are suitable for most inspection problems. For very high spatial resolution, piezoelectric driven micro-translators may be used. These can provide sub-micron sample translation with high accuracy. Sample scanning can become impracticable for the inspection of large areas because of the long time taken to gather all the necessary data.

Further reading

Aamodt, L.C., Murphy J.C. and Parker, J.C., 'Size consideration in the design of cells for photoacoustic spectroscopy', *J. Appl. Phys.*, **48** (1977) 927–33.

Boccara, A.C., Fournier, D. and Badoz, J., 'Thermo-optical spectroscopy: Detection by the "mirage effect"', *Appl. Phys. Lett.*, **36** (1980) 130–3.

Jackson, W. and Amer, N., 'Piezoelectric photoacoustic detection: Theory and experiment', *J. Appl. Phys.*, **51** (1980) 3343–53.

Kanstad, S.O. and Nordal, P.E., 'Photoacoustic and photothermal technique for powder and surface spectroscopy', *Appl. Sur. Sci.*, **6** (1980) 372–91.

Mandelis, A. and Zver, M.M., 'Theory of photopyroelectric spectroscopy of solids', *J. Appl. Phys.*, **57** (1985) 4421–30.

Miranda, L.C.M., 'Photothermal displacement spectroscopy of solids: theory', *Appl. Opt.*, **22** (1983) 2882–86.

Rosencwaig, A. and Gersho, A., 'Theory of the photoacoustic effect in solids', *J. Appl. Phys.*, **47** (1976) 64–69.

Rosencwaig, A., Opsal, J., Smith, W.L. and Willenborg, D.L., 'Detection of thermal waves through optical reflectance', *Appl. Phys. Lett.*, **46** (1985) 1013–15.

Vickery, I.P., 'Photoacoustic impulse response testing (PIRT) as a method for the non-destructive evaluation of the spectral and spatial depth profiling of solid samples', *Prog. Anal. Spectrosc.*, **11** (1988) 1–52.

5

Transient thermal phenomena

5.1 Propagation of heat pulses in solids

Heat conduction in a solid is a diffusion-like process where the thermal energy is transported by lattice vibrations or, in metals, by free electrons. The diffusion of thermal energy in a periodically heated solid was shown, in Chapter 2, to be described satisfactorily in terms of thermal waves. In this chapter we show that transient thermal effects can also be understood by making use of the thermal wave responses of a solid.

The reason for considering transient effects is that a number of important photothermal techniques employ pulsed, rather than cw, laser heating. For these techniques, information about the sample under investigation is obtained by the analysis of the thermal transient that follows the pulsed heating. Whilst it is not essential to develop this analysis by the thermal wave approach, considerable insight is gained by recognizing the links between transient responses and those that would be obtained in a thermal wave study. Formally, a thermal wave solution to a problem is a frequency-domain solution whilst a transient solution is a time-domain solution and the two are linked by a Laplace inversion. A time-domain response $T(t)$ may be obtained from a frequency-domain response using a Fourier series expansion:

$$T(x,t) = \sum_n a(\omega_n) T(x,\omega_n,t) \tag{5.1}$$

in which the component $T(x, \omega_n, t)$ is a plane thermal wave of angular frequency ω_n propagating in the x-direction (equation (2.22)) which is used as a basis function for the expansion:

$$T(x,\omega_n,t) = \frac{Q_0}{2k\sigma}\exp(-\sigma x)\exp(\mathrm{j}\omega_n t) \tag{5.2}$$

and where $a(\omega_n)$ is a measure of the strength of this component in the transient concerned.

The spectrum of thermal wave modulation frequencies is continuous and the Fourier series expansion (5.1) can be replaced by an integral transform. The resulting integrals can often be easily evaluated in the complex plane by the Laplace inverse transform method. The Laplace transform of a function $f(t)$ and its inverse are defined by the following equations:

$$f(s) = L\{f(t)\} = \int_0^\infty \exp(-st)f(t)\mathrm{d}t \tag{5.3}$$

$$f(t) = L^{-1}\{f(s)\} = \frac{1}{2\pi\mathrm{j}}\int_{c-\mathrm{j}\infty}^{c+\mathrm{j}\infty}\exp(st)f(s)\mathrm{d}s \tag{5.4}$$

where $t > 0$ and the symbols L and L^{-1} are used to denote the Laplace transform and its inverse respectively. The integration in equation (5.4), Bromwich's integral formula, is along a line $c + \mathrm{j}y$ in the complex plane, with c chosen to be to the right of all singularities but otherwise arbitrary. These transforms provide a simple route for obtaining the transient solutions to problems that have already been solved in the frequency-domain. For photothermal studies, this means that the thermal wave responses obtained in Chapter 2 can be utilized to derive the corresponding transient solutions. If we set $s = \mathrm{j}\omega$ we can then write the following general expression for the Laplace transformed surface temperature $T(s)$:

$$T(s) = Q(s)\,Z(s) \tag{5.5}$$

where $Q(s)$ and $Z(s)$ are the Laplace transformed heat source and thermal transfer function for the sample respectively. The time-domain surface temperature is simply:

$$T(t) = L^{-1}\{Q(s)Z(s)\} \tag{5.6}$$

To illustrate the above approach we will consider the transient thermal response of a semi-infinite solid to a Dirac delta surface heating pulse. This is an important example as the excitations produced in many pulsed laser experiments are modelled well by a Dirac delta function. A Dirac delta thermal impulse on the surface plane of the solid, $x = 0$, at time $t = 0$ is equivalent to a continuous broad frequency spectrum of equal energy thermal wave components that are all in phase only when $t = 0$ at $x = 0$. Details of the diffusion of the heat from the surface into the body of the solid can be understood in terms of the propagation of these thermal waves into the solid away from the surface. The different frequency components travel at different speeds (equation (2.24)) and suffer different amounts of attenuation (equation (2.23)). The temperature at a depth x and at a time t after the pulse excitation can be obtained by a simple addition of the thermal wave amplitudes at (x,t):

$$T(x,t) = \sum_{n=0}^{\infty} \frac{Q_0}{2k\sigma_n} \exp(-\sigma_n x)\exp(j\omega_n t) \tag{5.7}$$

which becomes:

$$T(x,t) = \frac{1}{\pi}\int_0^{\infty} \frac{Q_0}{2k\sigma(\omega)} \exp[-\sigma(\omega)x]\exp(j\omega t)\,d\omega$$

$$= \frac{1}{2\pi j}\int_0^{\infty} \exp(st)\frac{Q_0}{k\sigma(s)} \exp[-\sigma(s)x]ds$$

$$= Q_0 L^{-1}\left[\frac{\exp(-\sigma x)}{2k\sigma}\right] \tag{5.8}$$

demonstrating the role of the inverse Laplace transform in obtaining a transient response from a known frequency response. Setting $\sigma = \sqrt{(s/\alpha)}$ and evaluating the inverse transform (using one of the standard expressions tabulated in Table 5.1) leads to the solution corresponding to a Dirac delta pulse:

$$T(x,t) = \frac{Q_0}{2\sqrt{\pi\rho ckt}} \exp(-x^2/4\alpha t) \tag{5.9}$$

The general behavior of this function is shown in Fig. 5.1. The surface temperature $T(0,t)$ is inversely proportional to the square root of time and the thermal effusivity, $\sqrt{(\rho ck)}$. Beneath the surface the temperature $T(x, t)$ decays to 1/e of the surface magnitude in a

Table 5.1 Laplace transforms used in thermal transient analysis

$$v(s)=\int_0^\infty e^{-st}v(t)\,dt \quad s=j\omega \quad q=\left(\frac{s}{\alpha}\right)^{1/2}$$

$v(s)$	$v(t)$
I/s	1
I/q	$\sqrt{\alpha/\pi t}$
e^{-qx}	$\dfrac{x}{2\sqrt{\pi\alpha t^3}}e^{-x^2/4\alpha t}$
$\dfrac{e^{-qx}}{s}$	$\operatorname{erf} x\left(\dfrac{x}{2\sqrt{\alpha t}}\right)$
$\dfrac{e^{-qx}}{q}$	$\left(\dfrac{\alpha}{\pi t}\right)^{1/2}e^{-x^2/4\alpha t}$
$\dfrac{e^{-qx}}{sq}$	$2\sqrt{\alpha t}\ \operatorname{ierf} c\left(\dfrac{x}{2\sqrt{\alpha t}}\right)$
$\dfrac{e^{-qx}}{q(q+h)}$	$\alpha e^{hx+\alpha t h^2}\operatorname{erf} c\left(\dfrac{x}{2\sqrt{\alpha t}}+h\sqrt{\alpha t}\right)$
$\dfrac{e^{-qx}}{sq(q+h)}$	$\dfrac{2}{h}\left(\dfrac{\alpha t}{\pi}\right)^{1/2}e^{-x^2/4\alpha t}-\left(\dfrac{1+hx}{h^2}\right)\operatorname{erf} x\left(\dfrac{x}{2\sqrt{\alpha t}}\right)+\dfrac{1}{h^2}e^{hx+\alpha t h^2}\operatorname{erf} x\left(\dfrac{x}{2\sqrt{\alpha t}}+h\sqrt{\alpha t}\right)$

Where α and x are real and positive and h is real.

$$\operatorname{erf}(x)=\frac{2}{\sqrt{\pi}}\int_0^x e^{-u^2}\,du, \quad \operatorname{erf} x(x)=1-\operatorname{erf}(x) \text{ and}$$

$$\operatorname{ierfx}(x)=\frac{e^{-x^2}}{\sqrt{\pi}}-x\operatorname{erf} x(x)$$

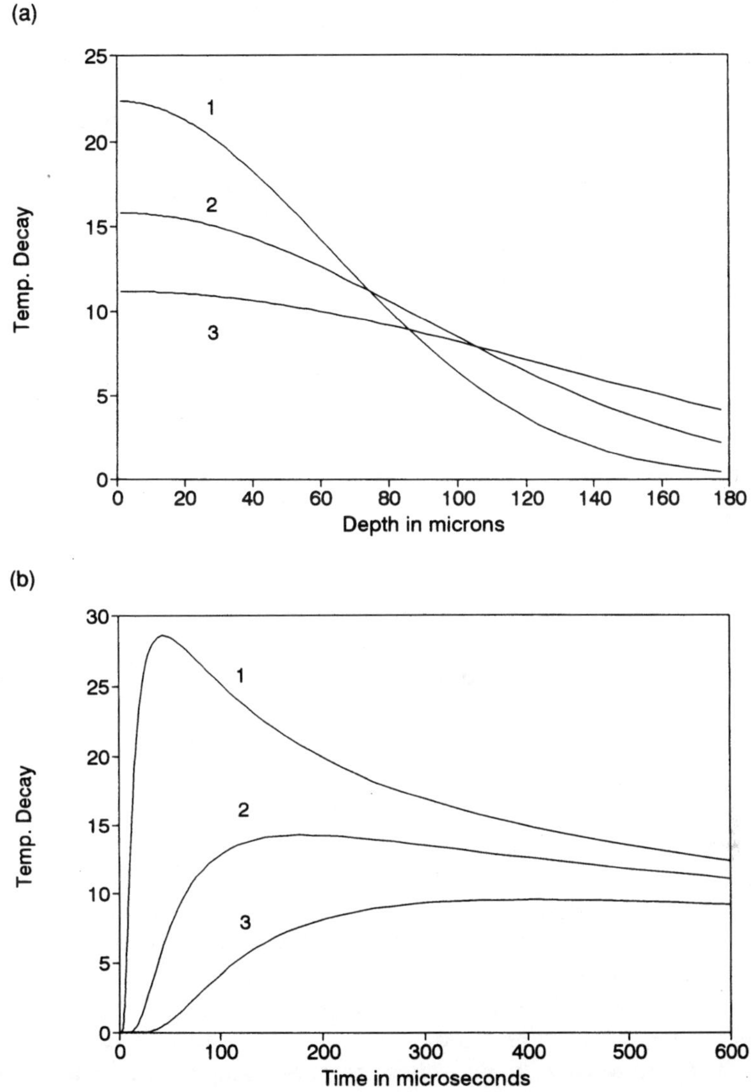

Figure 5.1 *Thermal transients resulting from surface pulse heating calculated using equation (5.9), and the thermal properties of stainless steel, shown (a) as a function of depth (1) 0.2, (2) 0.4 and (3) 0.8 ms after the pulse and (b) as a function of time at depths (1) 30, (2) 60 and (3) 90 μm.*

distance $x = 2\sqrt{(\alpha t)}$. This distance is called the effective thermal diffusion length μ_e in analogy with the thermal diffusion length of monochromatic thermal waves. It defines the depth of penetration of transient thermal energy.

The thermal wave interpretation of the development of the thermal transient is that the short time behavior is dominated by the fastest propagating, high frequency components of the pulse and the longer time behavior by the slower low frequency components. Because the high frequency components are heavily attenuated as they propagate, a sample can be considered to act as a low pass thermal wave filter, with only the slow moving low frequency components reaching deep into the sample.

5.2 Effects at interfaces

A plane thermal pulse arriving at an interface between two materials which have different thermal properties will be partly reflected and partly transmitted. The reflection coefficient for the pulse is identical with that already introduced in Chapter 2 for thermal waves:

$$R_b = \frac{1-b}{1+b} \qquad b = \sqrt{(\rho ck)_2/(\rho ck)_1} \tag{5.10}$$

This is because the reflection and transmission coefficients for a single interface between two materials are frequency independent and affect all the Fourier components of a pulse equally. This is not the case, however, when an additional layer or thermal contact resistance separates the two materials. The thermal wave reflection coefficients of thin layers and contact resistances (equation (2.41)) and (equation (2.44)), discussed in Chapter 2, are both frequency dependent, decreasing with decreasing frequency, Figs 2.11 and 2.13. Because of this, the reflection coefficient of a thin layer or contact resistance will vary throughout a transient experiment. At short times it will be high for the high frequency components but it will decrease at longer times where only the low frequency components remain. This behavior of the reflection function has important implications for the detectability of subsurface interfaces by transient photothermal techniques.

5.3 Transient thermal wave interference

Monochromatic thermal waves generated at a thin surface coating propagate in the coating causing constructive and destructive interference. The interference results from the combination of the reflected wave trains from the bottom and top surfaces of the coating. An expression for the surface temperature resulting from a plane pulsed surface heat source (optical) I_0 can be obtained from the Laplace inversion of the expression (2.38) for the periodic temperature at a coating surface:

$$T(t) = L^{-1}\left\{\frac{I_0[1-R(\lambda)]T_g}{4k_c\sigma_c}\left[\frac{1+R_b\exp(-2\sigma_c L_c)}{1-R_g R_b\exp(-2\sigma_c L_c)}\right]\right\} \tag{5.11}$$

Since the sample to gas reflection coefficient R_g is near unity and T_g is nearly two, the above expression can be simplified to:

$$T(t) = L^{-1}\left\{\frac{I_0[1-R(\lambda)]}{2k_c\sigma_c}\left[1+2\sum_{n=1}^{n=\infty}R_b^n\exp(-2\sigma_c nL_c)\right]\right\} \tag{5.12}$$

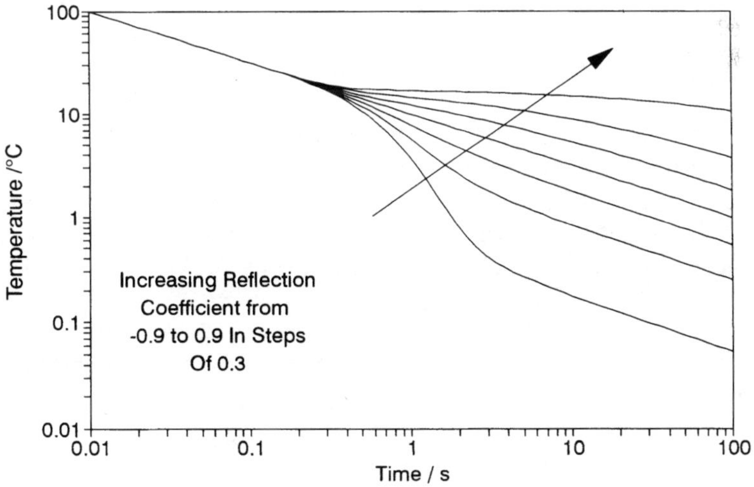

Figure 5.2 *Evolution of surface temperature with time for a layer–substrate system heated by a plane pulsed surface heat source. The layer is 1 mm thick and of a 10^{-6} m^2 s^{-1} diffusivity material.*

from which the solution for the sample surface temperature follows by using the Laplace inverse transforms in Table 5.1 term by term:

$$T(t)=\frac{I_0[1-R(\lambda)]}{2\sqrt{\pi\rho_c c_c k_c t}}\left[1+2\sum_{n=1}^{n=\infty}R_b^n\exp(-\frac{n^2L_c^2}{\alpha t})\right] \tag{5.13}$$

This expression is valid where the optical energy is absorbed at the surface and where the coating and substrate thermal properties are homogeneous. Figure 5.2 is a logarithmic plot of the decay of surface temperature with time calculated for coatings having a range of different thermal wave reflection coefficients. At short times the surface temperature decays as the inverse square root of time. At these times the pulse has not reached the film–substrate interface and the surface temperature response is determined solely by the thermal properties of the coating layer. As time increases, deviations from this $t^{-1/2}$ decay rate occur due to the thermal mismatch between coating and substrate, resulting in multiple reflections of the thermal wave components of the pulse within the coating layer. The significance of the summation index 'n' in equation (5.13) is now apparent. It shows the number of round trips a thermal pulse has made in its contribution to the surface temperature. Each added pulse reflection has a decreasing effect on the surface temperature amplitude due its rapid attenuation with distance travelled. The sign of the thermal wave reflection coefficient (indicating whether the substrate has a higher or lower thermal effusivity than the coating) and its magnitude determine the degree of deviation from the $t^{-1/2}$ characteristic. The onset of this deviation is determined by the characteristic propagation time of the heat pulse in the coating: $t_p = L_c^2/\alpha_c$. At long times the surface layer becomes thermally thin and the temperature decay follows an inverse root time decay as before but at a new amplitude determined by the substrate thermal effusivity.

The dependence of temperature–time data on surface optical properties can be eliminated by plotting the slope of the logarithmic temperature verses time:

$$\text{slope}(T)=\frac{\mathrm{d}'\log(T(t))}{\mathrm{d}'\log(t)}=\frac{t}{T}\frac{\mathrm{d}T}{\mathrm{d}t} \tag{5.14}$$

Figure 5.3 shows the functional variation of slope with time for various layer/substrate reflection coefficients. The deviation from

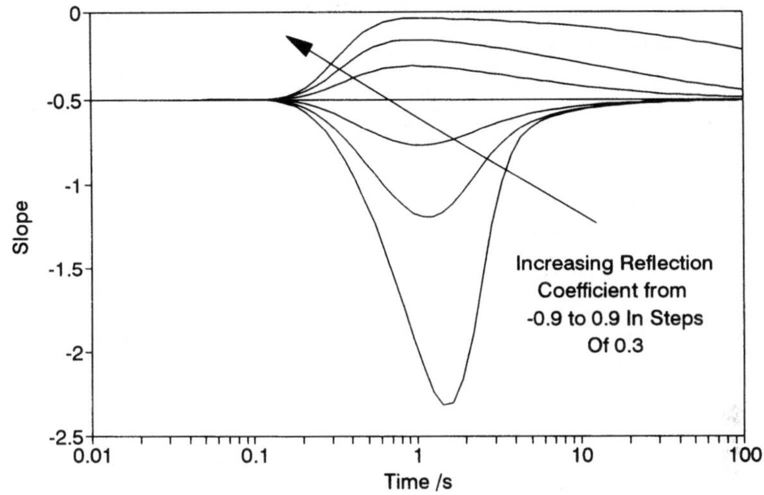

Figure 5.3 *Variation of the logarithmic gradient of temperature decay curves of Fig. 5.2 with time.*

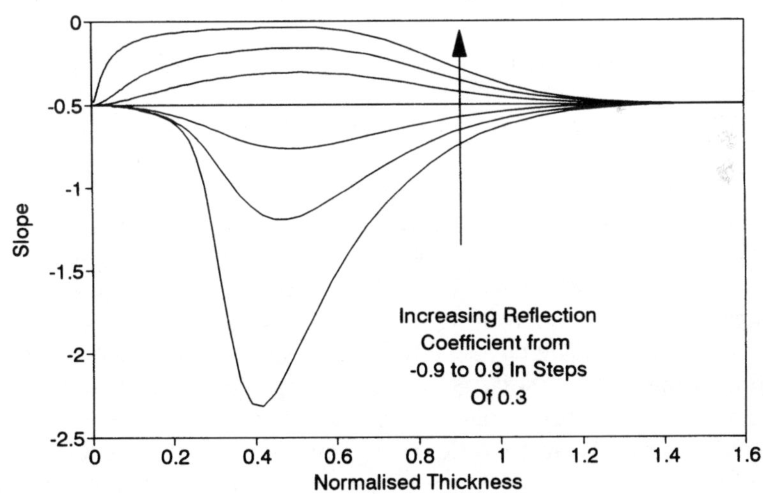

Figure 5.4 *Variation of the logarithmic gradient of temperature decay curves of Fig. 5.2 with normalized thermal coating thickness,* L/μ_e.

the slope of –1/2 is now clearly apparent. Positive reflection coefficients result in an increase of the slope value (more positive) and negative coefficients lead to a decrease in the slope value. At short and long times the slope values become independent of the thermal properties of the coating/substrate and of time, going to a limiting value of –1/2.

A normalized thermal transient interference graph may be produced by plotting slope variation against normalized effective layer thickness, $L_c/\mu_e = L_c/2\sqrt{(\alpha_c t)}$, as shown in Fig. 5.4. The similarity of this figure with Fig. 2.8, showing the variation of phase of a thermal wave with the thermal thickness of a coating should be noted. It can be seen that the transient thermal wave interference effects are evident up to 1.3 μ_e with a maximum value at around 0.3–0.4 μ_e. This transient slope interference characteristic is more asymmetric than that for the thermal wave phase. This asymmetry may be beneficial for layer thickness measurements or interface characterization.

5.4 Laser generation of a heat pulse

The use of laser sources in transient photothermal studies necessitates the development of a solution of the heat conduction equation which takes account of laser spot size, finite optical absorption and temporal variation.

Optical spot size effects can be determined by evaluating a convolution integral over the spatial extent of the heated region. It is common to model a laser spot by a Gaussian with 1/e radius r_B. It was shown in Section 4.1.1 that the size of the heat source plays an important role in controlling the degree of heat diffusion. There are conditions under which heat diffusion must be considered to be three-dimensional (for a small spot size) and others where simple one-dimensional thermal transport occurs (for a large spot). In the transient case these conditions are determined by the relative magnitudes of r_B and μ_e. Figures 5.5 and 5.6 show the surface temperature and slope variation with time at the center of the laser heating beam of 1/e radius 1 mm. At short times the temperature decays as $t^{-1/2}$. At these times the laser spot size appears to be much larger than the effective range of heat diffusion, as determined by the transient thermal diffusion length. The range of heat diffusion increases with time and eventually becomes larger than the laser spot size, the sur-

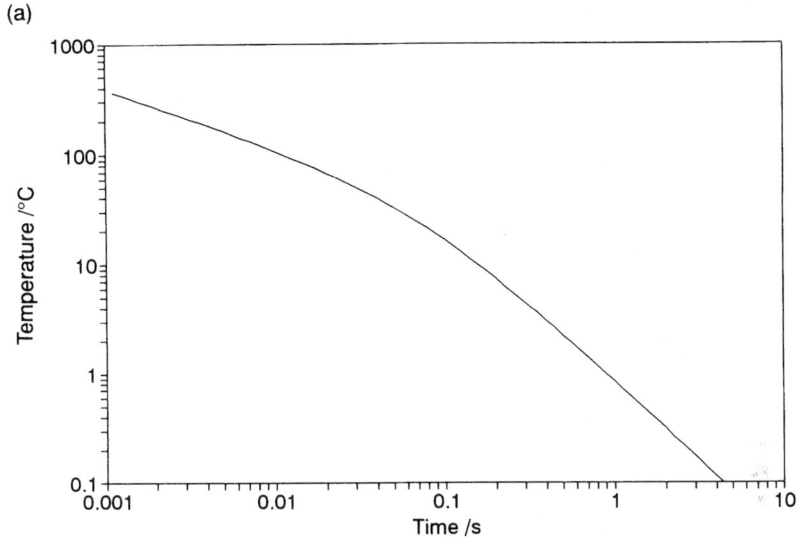

Figure 5.5 *Variation of the surface temperature with time at the center of a spot heated by a laser pulse of l/e radius 1 mm, material: steel.*

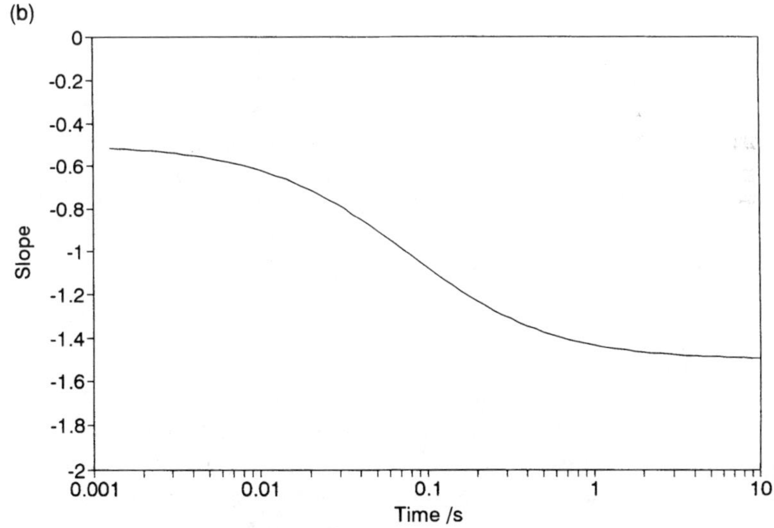

Figure 5.6 *Slope variation with time of temperature profile in Fig. 5.5.*

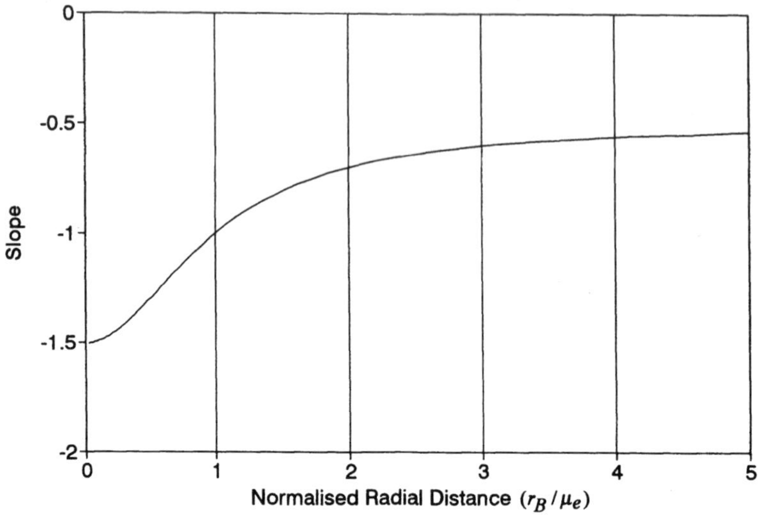

Figure 5.7 *Slope variation with normalized laser spot radius* r_B/μ_e, *of temperature profile in Fig. 5.5.*

face temperature then decaying at a much faster rate, $t^{-3/2}$. At these long times the radial heat spreading is sufficient to affect the center of the laser spot. By plotting the slope variation with respect to the normalized laser spot size, r_B/μ_e, it can be seen from Fig. 5.7 that the measured temperature at the beam center can be regarded as that due to a plane pulsed heat source, producing a $t^{-1/2}$ decay with time, provided that $r_B > 4\mu_e$. This result is the equivalent of that in Section 4.1.1 for a sample surface heated by a periodically modulated Gaussian profile laser beam.

5.5 Finite absorption effects

The finite absorptivity and transparency of materials will affect their photothermal transient responses. For thermal waves, in the frequency-domain, it was found that the amplitude dependence on frequency was stronger in translucent materials. In addition, it was noted that the phase angle variation was between two limits: photothermally opaque and transparent. A similar result is also obtained in the time and slope evolution of the transient surface temperature. By using equation (3.23) and performing the inverse Laplace transform we can obtain the following expression for the

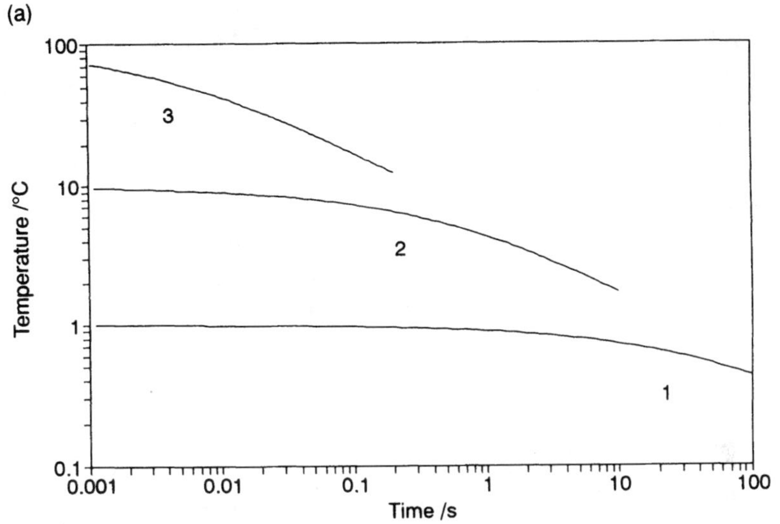

Figure 5.8 *Surface temperature variation with time for translucent semi-infinite samples, optical absorption coefficients (1) 100, (2) 1000 and (3) 10 000 m^{-1}, subjected to plane pulsed heating.*

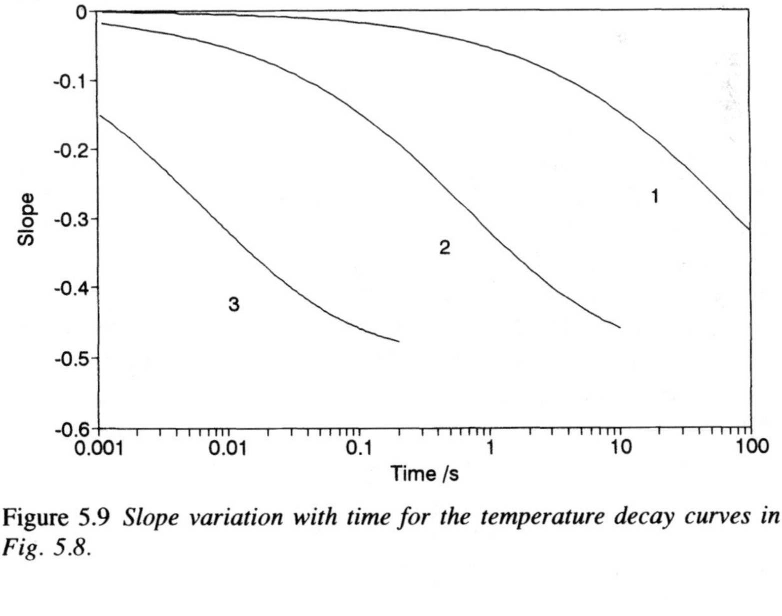

Figure 5.9 *Slope variation with time for the temperature decay curves in Fig. 5.8.*

surface temperature variation of a semi-infinite optically translucent material subjected to plane pulsed optical excitation:

$$T(t) = I_o[1 - R(\lambda)]\frac{\alpha}{2k}\exp(\alpha\beta^2 t)\mathrm{erfc}\left(\beta\sqrt{\alpha t}\right) \qquad (5.15)$$

In deriving this equation we have assumed an exponential absorption of optical radiation by the sample. Figures 5.8 and 5.9 show the surface temperature and slope variation with time for materials with optical absorption coefficients of 100, 1000 and 10 000 m^{-1}. At short times, where the effective transient thermal diffusion depth is small compared with the optical penetration depth, the surface temperature is dominated by the optical absorption depth and its temporal variation will be nearly independent of time. At very long times, where the transient thermal diffusion depth is greater than the optical absorption depth, the surface temperature will decay as $t^{-1/2}$. A transition region exists between these two limits as shown in Fig. 5.9.

5.6 Pulse shape effects

The temporal shape of the applied heat source can lead to a significant modification of sample surface temperature evolution. For example the excitation source may be such that it remains uniform in time (step pulse) or is of a finite duration or rises and decays according to a predetermined function (exponential, triangular etc.). Whatever the temporal variation of the heat source, equation (5.1) can be used to derive the surface temperature variation. All that is required is to obtain the Laplace transform of the excitation function, multiply it by the system thermal transfer function and compute the Laplace inversion. As an example let us consider the two cases cited above: a step on pulse and a step pulse of finite duration applied to a semi-infinite sample. The surface temperature for these two cases is obtained from:

step pulse

$$T(t) = L^{-1}\left\{\frac{Q_o}{s}\frac{1}{k\sigma}\right\} = \frac{2Q_o}{k}\sqrt{\frac{\alpha t}{\pi}} \qquad (5.16)$$

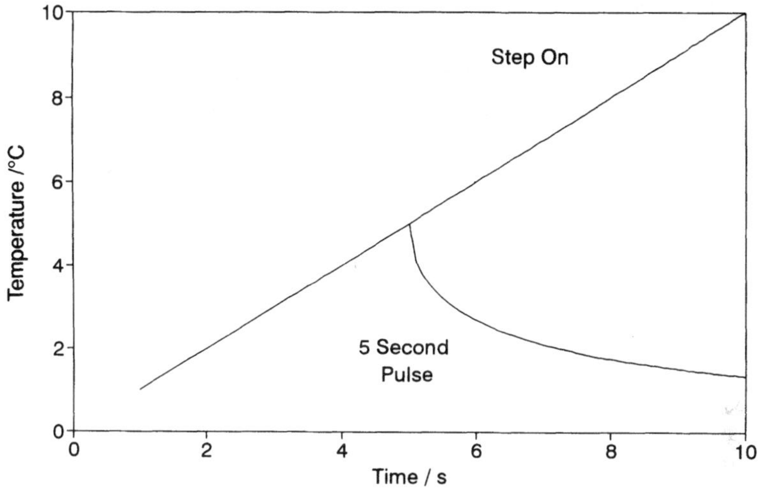

Figure 5.10 *Surface temperature variation resulting from a plane surface step pulse excitation of a semi-infinite material and also that resulting from a 5 s duration pulse.*

finite pulse

$$T(t) = F(t) - \delta(t)F(t-\tau) \qquad \begin{matrix} \delta(t) = 0 & t \leq \tau \\ \delta(t) = 1 & t > \tau \end{matrix} \tag{5.17}$$

where $F(\ )$ represents the temperature response due to a step pulse excitation.

Figure 5.10 illustrates the temporal variation for these two forms of excitation. For a step on pulse excitation the surface temperature varies as the square root of time and inversely with the thermal effusivity of the sample. The effect of a finite pulse duration is a decay of the surface temperature when the excitation is switched off. The temperature response reduces to that of a pulsed excitation source for narrower pulse widths. The pulsed response of the heated sample can be obtained from the first derivative of the step pulse temperature response function.

5.7 Convective heat loss effects

Some pulsed heating techniques produce considerably higher sample surface temperature changes than their cw counterparts. This raises the possibility of significant heat loss by convection, or even by radiation. The magnitude of convective heat loss depends on the temperature difference between the heated surface and its surrounding gas or fluid (usually air), see Section 2.1:

$$q_{cv} = h(T_s - T_\infty) \tag{2.2}$$

The importance of this term is dependent on the relative magnitude of the conductive heat loss term at the surface:

$$q_{cd} = -k\left.\frac{\partial T}{\partial x}\right|_{surface} \tag{5.18}$$

Convective heat loss can be particularly important in experiments on low conductivity materials for two reasons:

1. Conductive loss is directly proportional to thermal conductivity and will be small in low conductivity materials.
2. Photothermally induced surface temperature rises vary inversely with conductivity and, consequently, it is possible in an experiment to create a much larger surface temperature change in a low conductivity material, using a specific optical energy. This leads directly to an increase in the magnitude of the convective loss term.

In an extreme case, the convective term may far exceed the conductive term and a cooling transient will exhibit the simple exponential time dependence associated with Newton's law of cooling. In practice it is advisable to model an experiment to determine the importance of the two terms.

An expression for the transient surface temperature variation of a semi-infinite sample, heated at its surface by a pulsed plane heat source, can be obtained by solving the one-dimensional heat diffusion equation with the boundary conditions:

$$-k\frac{dT}{dx}\Big|_{x=0} = (1-r_0)Q - h_e T(0) \qquad T(x=\infty, t) = 0 \tag{5.19}$$

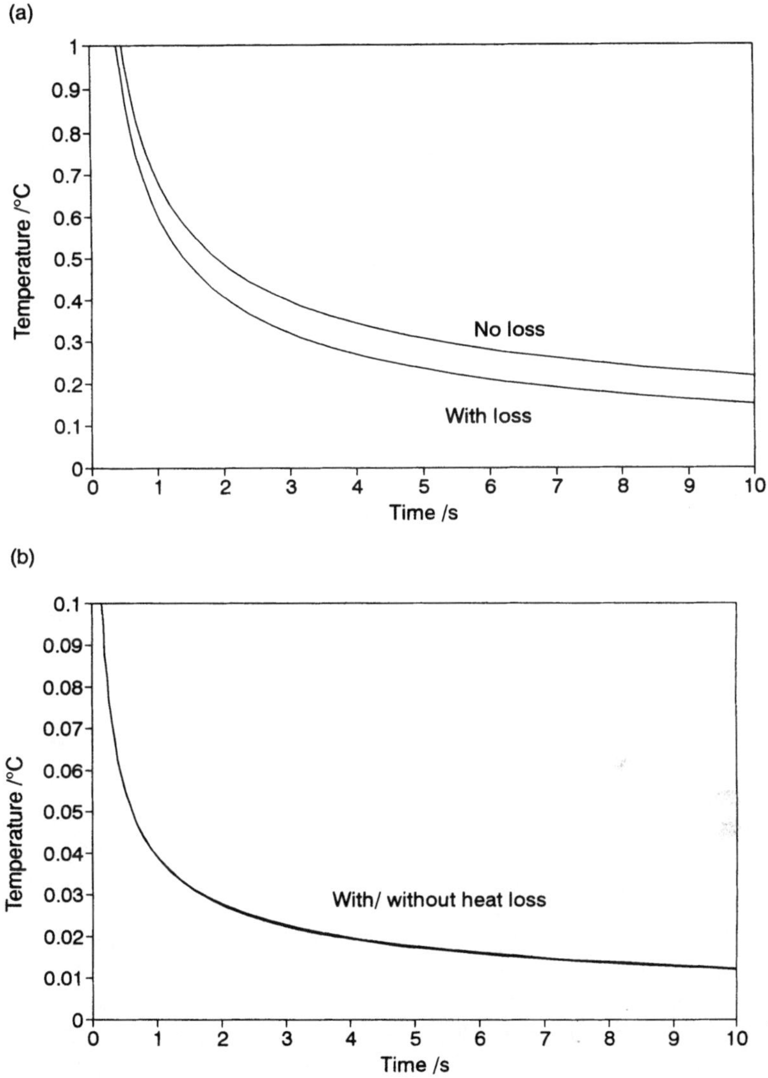

Figure 5.11 *Transient response to pulsed heating of (a) resin and (b) stainless steel. The two curves were computed with and without the inclusion of a convective heat loss term.*

The solution leads to the following expression:

$$T(0,t) = [1 - R(\lambda)]Q\left[\left(\frac{\alpha}{\pi t}\right)^{1/2} - H\alpha \exp(H^2\alpha t)\text{erfc}(H\sqrt{\alpha t})\right] \quad (5.20)$$

where Q is the pulse energy and $H = h_e/k$, an effective heat transfer coefficient. H accounts for the convective and possible radiative heat loss from the sample surface:

$$h_e = h_{cv} + h_{rd} \approx h + 4\varepsilon\sigma_{SB}T_0^3 \quad (5.21)$$

T_0 is the sample temperature. It can be readily shown that, for a sample at or near room temperature, the radiative contribution to the surface heat loss is much smaller than that due to convection.

Examples of modelled transient responses are shown in Fig. 5.11 for (a) a low thermal conductivity resin and (b) stainless steel. Curves are shown computed with and without the additional convective heat loss term. It is clear that the long time transient response of the resin is dominated by convection whilst the inclusion of convection has no effect on the transient response of stainless steel.

Further reading

Arpaci, V.S., *Conduction heat transfer*, Addison Wesley Publishing (1966).

Balageas, D.L., Krapez, I.C. and Cielo, P., 'Pulsed photothermal modelling of layered materials', *J. Appl. Phys.*, **59** (1986) 348–57.

Carslaw, H.S. and Jaeger, J.C., *Conduction of heat in solids*, Oxford University Press (1959).

Murphy, J.C., Aamodt, L.C. and Spicer, J.W.M., 'Principles of photothermal detection in solids', in *Principles and perspectives of photothermal and photoacoustic phenomena* (ed. A. Mandelis), Elsevier Science, New York (1992).

6

Photoacoustic spectroscopy

The development of photoacoustic spectroscopy in the 1970s provided much of the groundwork for the many and varied manifestations of photothermal science in use today. Photoacoustic spectroscopy is a branch of absorption spectroscopy in which the sample heating caused by the selective absorption of optical energy is monitored. Whilst there are many different mechanisms of optical absorption, many of these processes eventually result in the absorbed optical energy being dissipated as heat. This heat produces a variety of physical changes in a sample and its surroundings that can be measured as a means of assessing optical absorption. The physical processes that may contribute to the photoacoustic effect are shown schematically in Fig. 6.1.

Photoacoustic spectroscopic techniques are now well established and a number of leading spectrometer manufacturers offer photoacoustic cell accessories. Photoacoustic spectroscopy has a number of attractive features which account for its adoption by spectroscopists:

1. requires minimal sample preparation;
2. enables measurements of highly absorbing and scattering media;
3. can be used on a broad range of materials (gases, liquids, solids, powders, gels, thin films);
4. can be used to determine a very wide range of absorption coefficient magnitudes – 10^{-3} to 10^{5} m^{-1};
5. non-contact, non-destructive and the potential for remote sensing;
6. spectral depth profiling;
7. operation in hostile environmental conditions;
8. a range of complementary photothermal detection methods;
9. an increase in S/N ratio with increase in input light power.

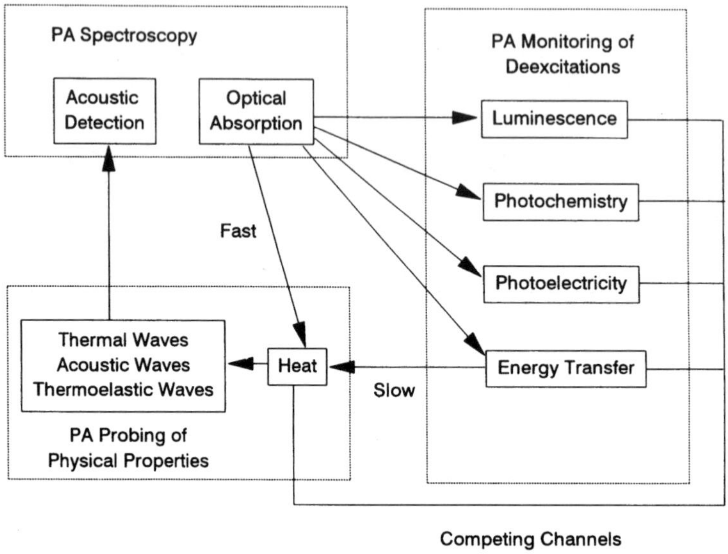

Figure 6.1 *Schematic diagram of the photoacoustic effect.*

In addition to obtaining spectra for identification purposes, the quantitative determination of optical properties is important for many materials applications and photoacoustic spectroscopy has much to offer in the field of quantitative spectroscopy.

The photoacoustic and other photothermal methods that are used for evaluating some of the common optical properties of solid materials are explained in this chapter. This is followed by examples of applications, taken from a wide range of areas, to illustrate the scope of this spectroscopic technique.

6.1 Photoacoustic spectrometers

Figs 6.2 and 6.3 are schematic diagrams of the two most common systems used for solid state photoacoustic spectroscopy in the ultraviolet to near infrared (UV–NIR) and the infrared (IR) spectral regions. The UV–NIR system employs: a high powered arc discharge lamp; a mechanical chopper, a scanning grating monochromator and a photoacoustic (PA) cell detection system. The PA signal is recorded as a function of the wavelength selected (typically

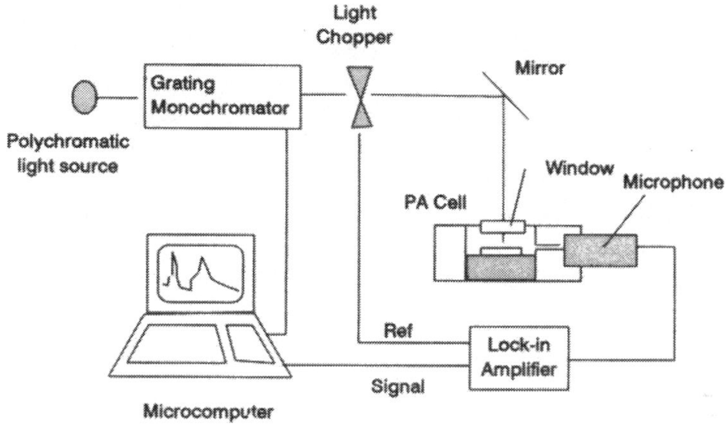

Figure 6.2 *Schematic diagram of an UV–visible photoacoustic spectrometer.*

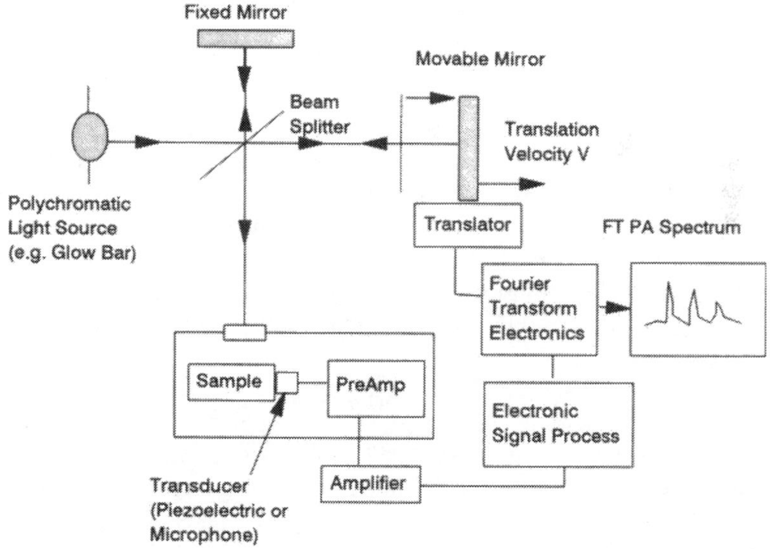

Figure 6.3 *Schematic diagram of a Fourier transform infrared photoacoustic spectrometer.*

10 nm bandwidth) by slowly scanning the monochromator. A spectrum is obtained from the ratio of the PA signal from the test sample to that from a black absorber (generally carbon black). This normalization eliminates the spectral variation of the illumination source. An absorption spectrum is obtained if the condition $\beta(\lambda)^{-1} > \mu_s(f)$ is satisfied at a particular wavelength. Examples of absorption spectra obtained in this way are presented later in this chapter.

For infrared spectroscopy the above system becomes unsuitable for several reasons. The available broadband IR sources have considerably lower power than ultraviolet–visible (UV–VIS) sources. Significant losses in incident power occur at the grating monochromator and at the input and exit slits. The need to cover the large spectral range of the IR necessitates a much faster spectral sampling method than that used in the UV–VIS. To overcome these problems the Fourier transform spectroscopy (FTS) technique has been widely adopted in combination with a number of the photothermal detection techniques, PA, optical beam deflection (OBD) and photo-pyroelectric (PPE). The advantages of FTS are that data at all spectral wavelengths emitted by the source are obtained simultaneously (Fellgett's advantage) and the optical throughput of the Michelson interferometer used in FTS is much larger than that of a dispersive spectrometer (Jacquiot's advantage). Two types of FTS are commercially available: one operating in a continuous scan mode and the other in a step-and-scan mode. In the continuous scan mode the moving mirror of the interferometer is scanned at a fixed velocity V. As this mirror moves, constructive and destructive interference occurs resulting in a sinusoidal modulation of beam intensity. The modulation of the external source is such that it depends on the mirror velocity and spectral wave number, $\bar{\nu}$:

$$I(\bar{\nu}, V) = 0.5B(\bar{\nu})[1 + \cos(2\pi 2\bar{\nu}V t)]$$

The term $B(\bar{\nu})$ accounts for the source spectral variation and the reflection efficiencies at the beam splitter. The imposed modulation on the source has a frequency $= 2V\bar{\nu}$. For the spectral region 400-4000 cm^{-1}, and a typical mirror velocity of 0.1 cm s^{-1}, the modulation frequency varies from 80 to 800 Hz. It is important to note that because the modulation frequency varies with the spectral wave number the depth of the region sampled (the thermal diffusion length) varies across the spectrum. This variation of the sampling depth presents a problem in the quantitative analysis. One solution is to obtain a ratioed spectrum with data recorded at different mirror speeds. An alternative solution, and the one now most generally used for quantitative studies, is to employ the step and scan FTS method. In the step–scan operation, beam modulation at a particular wave number is achieved by first moving the movable mirror of the interferometer to a particular location and then dithering the position of this mirror at this location. This imposes a phase modulation on the external source and decouples the dependences of the modulation frequency on mirror velocity and spectral wave number. Lock-in detection is then used to obtain the amplitude and phase of the modulated photothermal signal. By this method a fixed sampling depth is guaranteed across the whole spectral range.

6.2 Reflectivity measurements

The spectral variation of the surface reflectance of an optically opaque material can be obtained using a photothermal technique. The general expression for photothermal signal variation with wavelength can be written as:

$$S_s(\lambda) = F_s(\lambda)\eta_s(\lambda)[1 - R_s(\lambda)]I_0(\lambda) \tag{6.1}$$

in which $F_s(\lambda)$ is a parameter determined by the optical, thermal and geometric properties of the sample and detector (PA cell, PZT transducer, pyroelectric sensor) and the light modulation frequency; $I_0(\lambda)$ denotes the intensity variation of the exciting light source with wavelength; $\eta_s(\lambda)$ and $R_s(\lambda)$ are the non-radiative recombination efficiency and surface reflectivity of the medium under investigation respectively. To obtain $R_s(\lambda)$, a normalization of the signal must be performed using a spectrally flat absorber. A material commonly used as a reference sample for this normalization process is carbon black. Carbon black has a $\eta_s(\lambda)$ near unity, negligible reflectivity and the absence of absorption bands, particularly in the visible region of the spectrum. Dividing the sample signal by that from the reference sample allows the source intensity variation to be eliminated, resulting in a normalized signal which is dependent solely on reflectivity:

$$S_N(\lambda) = \frac{S_s}{S_r} \propto [1 - R_s(\lambda)] \tag{6.2}$$

assuming, as is often the case, that $\eta_s(\lambda)$ for the sample is also near unity.

6.3 Optical absorption coefficient determinations

The optical absorption coefficient can be determined using the PA, PPE, OBD and photothermal radiometry (PTR) techniques with either modulated (cw) or pulsed forms of optical excitation. We will concentrate here on the more common cw variety. Three methods can be employed if the sample is optically and thermally thick.

Case 1: Measurement at single wavelength

The photothermal signal's phase dependence on the light source modulation frequency can be used to determine the optical absorption coefficient. It was pointed out in Chapter 3 that the dependence of the normalized phase on frequency is given by the expression, (3.24):

$$\varphi(\beta, f) = -\tan^{-1}\left(1 + \frac{2}{\beta\mu}\right) \tag{6.3}$$

The normalization of the photothermal signal here is with respect to an optically opaque and thermally thick sample. This process removes phase variations due to the electronic components of the measurement system. As discussed in Chapter 3, the normalized phase of a translucent sample varies between 0 and –45°, the phase of 0° arising from an optically opaque sample and the phase of –45° coming from a transparent one. The optical absorption coefficient β may be determined from equation (6.3) provided that the thermal diffusivity of the sample is known with sufficient accuracy to calculate the value of the thermal diffusion length μ.

Case 2: Over a broad spectral region

The following procedure can be used to transform a recorded photothermal spectrum determined by the PA, OBD or PTR methods to a form that is independent of the optical wavelength dependent parameters $R_s(\lambda)$ and $\eta_s(\lambda)$. Assuming that the sample is translucent, the normalized signal amplitude (from equation (3.23)) can be expressed as follows:

$$S(\beta, f) = A_s C \left\{ \frac{\beta\mu}{[(\beta\mu + 1)^2 + 1]^{1/2}} \right\} \tag{6.4}$$

where C is a constant factor determined by the characteristics of the detection system employed and

$$A_s = \eta_s[1 - R_s(\lambda)]I_0(\lambda).$$

This equation has essentially two unknowns: CA_s and $\beta\mu$. The ratio of signals obtained at two different modulation frequencies eliminates CA_s:

$$\frac{S_1}{S_2} = \frac{\mu_1}{\mu_2} \sqrt{\left[\frac{(\beta\mu_2 + 1)^2 + 1}{(\beta\mu_1 + 1)^2 + 1} \right]} = g_s \tag{6.5}$$

By writing

$$m = \mu_1 / \mu_2 = \sqrt{f_2 / f_1}\ ,$$

the above equation can be solved for $\beta\mu_1$ in terms of g_s and m:

$$\beta\mu_1 = \frac{\left\{ \left(m - g_s{}^2\right) + \sqrt{\left[\left(m - g_s{}^2\right) + 2\left(m^2 - g_s{}^2\right)\left(g_s{}^2 - 1\right)\right]} \right\}}{g_s{}^2 - 1} \tag{6.6}$$

where $1 < g_s < m$. This can be repeated for all spectral wavelengths. Again the absolute value of the optical absorption coefficient can be determined, provided that the thermal diffusivity of the sample is known, to calculate the value of the thermal diffusion length μ_1.

Case 3: Thin film normalized with respect to an optically opaque sample

This method is applicable to an optically and thermally thin sample. It can be shown (equation (3.29)) that in these limiting conditions the photothermal signal for PA and OBD detection is directly proportional to the optical thickness:

$$S(\beta) = A_s C \beta L_s \quad (6.7)$$

By obtaining the ratio of this signal with respect to that when the sample is optically opaque the constant term C can be eliminated as the ratio depends only the product A_s and βL_s. The dependence on A_s is relaxed when measurements of this signal are taken near a high absorption region. Alternatively, a complete account of A_s is given by expression (3.33) which incorporates optical interference effects, as discussed in Chapter 3. For this measurement method a knowledge of the thin film thickness is required (and additionally the refractive index of the film and substrate) for absolute optical absorption coefficient determination.

6.4 Application 1 – metallic materials

For highly absorbing materials such as metals the non-radiative quantum efficiency is near unity and the normalized photothermal signal depends in a simple way on reflectivity, equation (6.2). Figs 6.4–6.6 show the normalized photoacoustic signal obtained from powdered copper and silver, mild steel and plasma sprayed copper. These qualitative spectral measurements illustrate the effectiveness of the photoacoustic technique in dealing with samples of types that would be difficult to study using other spectroscopic methods.

The determination of reflectivity spectra by the OBD and modulated optical reflectance (MOR) techniques is complicated by their signals being functions of additional wavelength dependent parameters. For OBD measurements this originates from the refractive index temperature dependence, $dn(\lambda)/dT$, and for MOR from the reflection coefficient $dR(\lambda)/dT$. For OBD measurements the signal dependence on $dn(\lambda)/dT$ of the gas or fluid above the sample can be eliminated by a normalization process. However, this is not possible for MOR measurements because $dR(\lambda)/dT$ is a sample property that would not be common to a reference sample. This result does not imply that $R(\lambda)$ cannot be obtained with an MOR system as $R(\lambda)$ can be obtained simply (and more directly) by monitoring the reflected light.

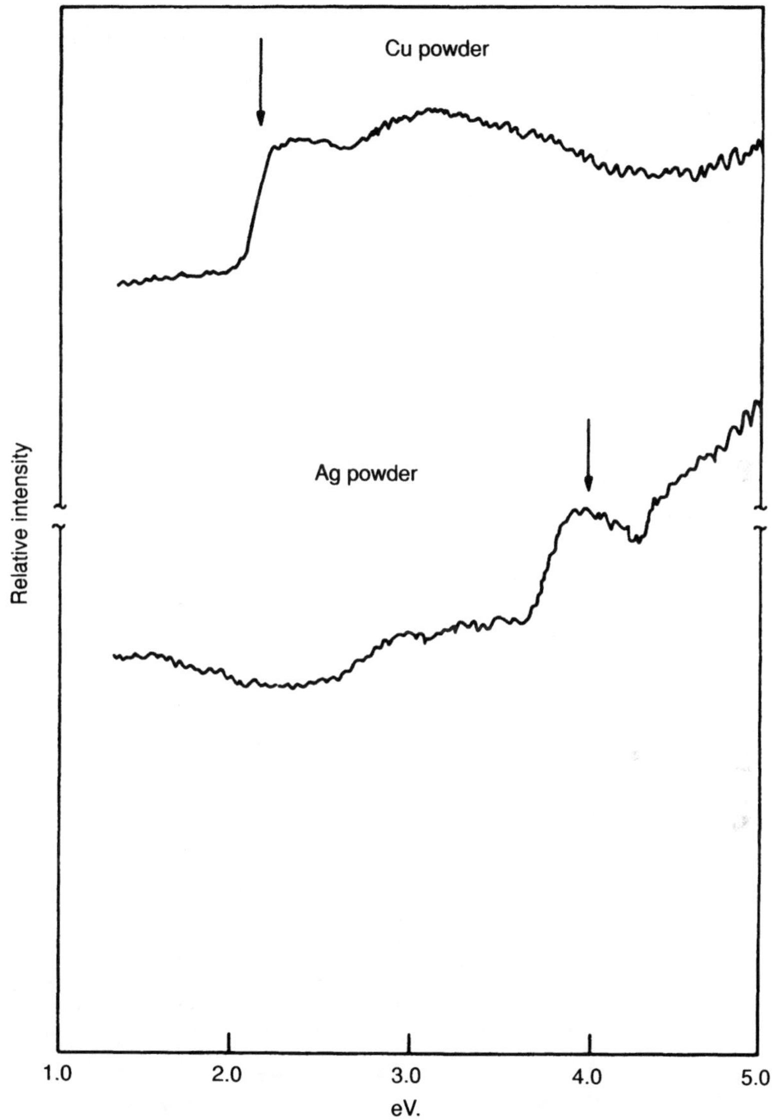

Figure 6.4 *Photoacoustic spectra of powdered copper and silver. (Reproduced with permission from A. Rosencwaig. Copyright 1975. American Chemical Society.)*

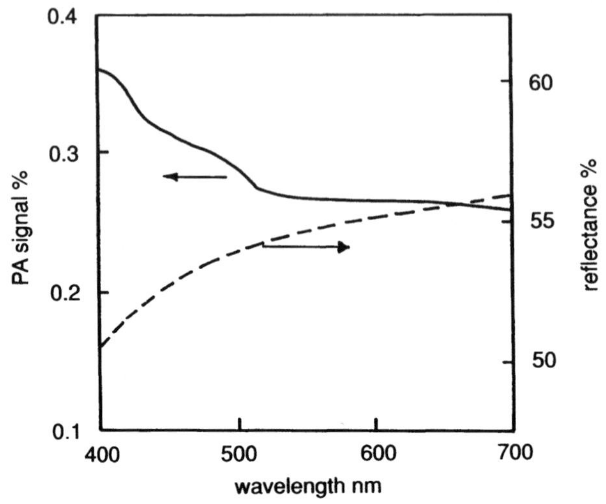

Figure 6.5 *Photoacoustic and reflectance spectra of mild steel (after Futamata et al., 1992).*

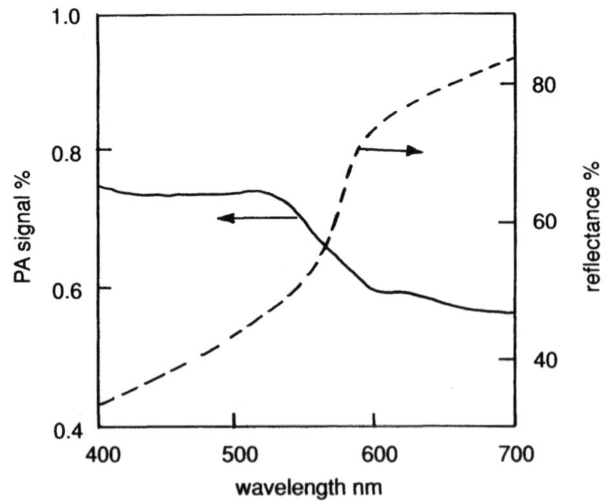

Figure 6.6 *Photoacoustic and reflectance spectra of copper plasma sprayed onto a mild steel substrate (after Futamata et al., 1992).*

6.5 Application 2 – semiconducting materials

Photothermal spectroscopy has been employed in numerous spectral characterizations of semiconducting materials in the forms of: amorphous layers; single crystals; compounds and multi-layer systems. Provided $\mu_\beta > \mu_s$ over the spectral illumination range required, spectroscopy can be performed using photothermal techniques. It has been shown that absorption spectra can be obtained readily from powdered semiconducting materials despite the strong light scattering in these materials. A good correlation is found

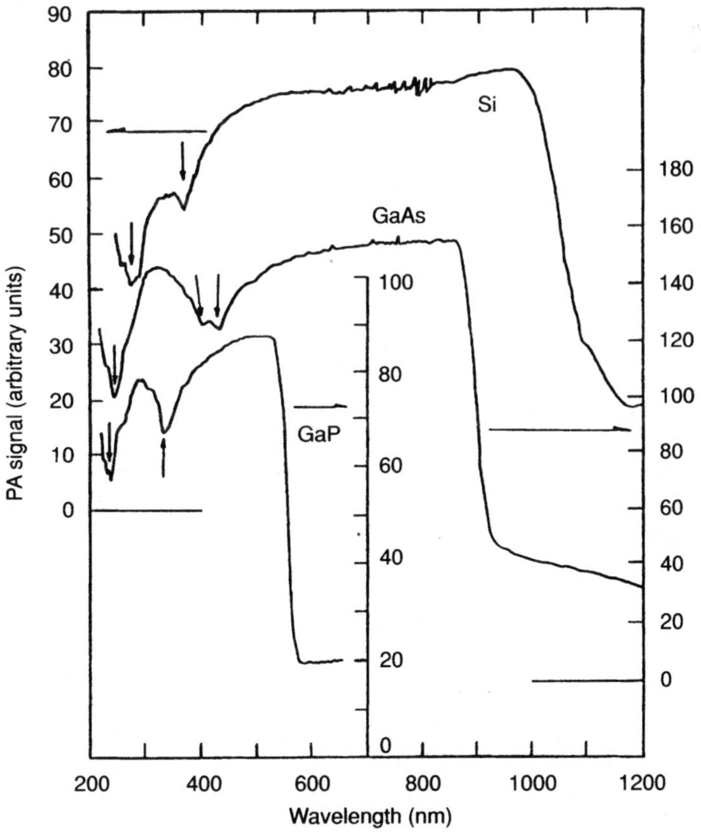

Figure 6.7 *Photoacoustic spectra of silicon, galium arsenide and gallium phosphide (Tokumoto et al., 1981: reproduced with permission).*

between photoacoustically measured band-gaps and those obtained by the diffuse reflectance measurements. Examples of studies of crystalline cubic semiconducting materials (Si, Ge, GaAs, GaP and InSb) covering the spectral range between 200 and 1200 nm at a frequency of 40 Hz are shown in Figs 6.7 and 6.8. Clear indications of the fundamental absorption edges are evident for Si, GaAs and GaP (1.12, 1.43 and 2.3 eV respectively) which are in good agreement with published values obtained by other techniques. The absorption edges from Ge and InSb are not present because they are beyond the instrument's spectral range. At photon energies above the band-gap, 'dips' in the PA amplitude are also evident. The origins of these absorption variations were found to be resonant absorptions via interband transitions. These absorption variations alter the sample reflectivity and hence the PA signal amplitude. In

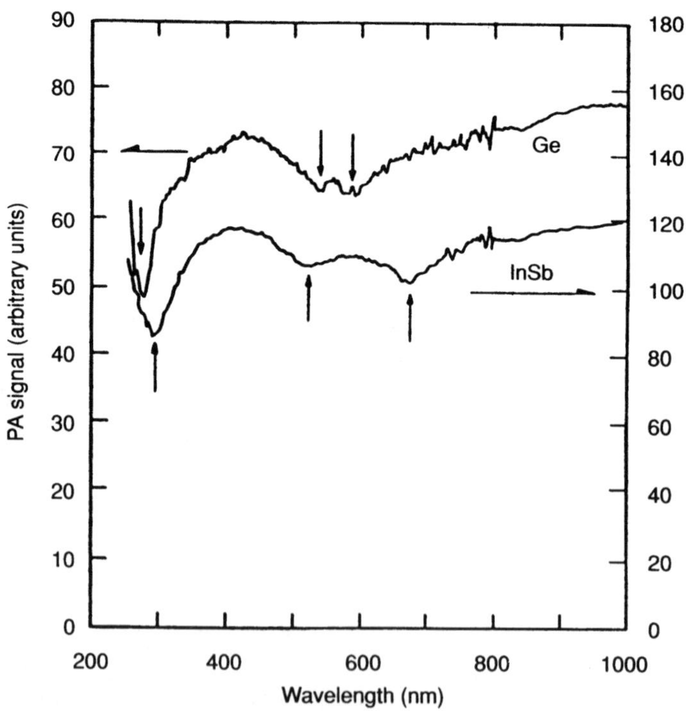

Figure 6.8 *Photoacoustic spectra of germanium and indium antiminide (Tokumoto et al., 1981; reproduced with permission).*

this high energy spectral region the PA signal depends only on the sample reflectivity, as discussed in Chapter 3.

The quantitative analysis of the photothermal signal and phase has been used by various researchers to obtain optical absorption coefficient values. Figures 6.9 and 6.10 demonstrate the use of the PA signal and phase spectral data to obtain, through the ratio method discussed above, the spectral variation of the optical absorption coefficient of crystalline GaSe. This data was found to be in good agreement with that determined by transmission measurements.

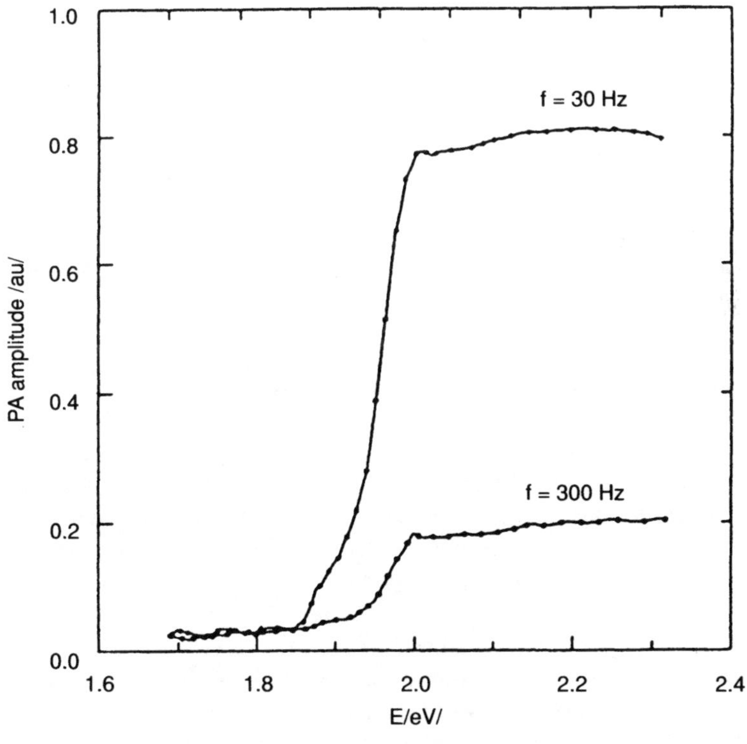

Figure 6.9(a) *Photoacoustic signal amplitude of gallium selenide (Todorovic et al., 1991; reproduced with permission).*

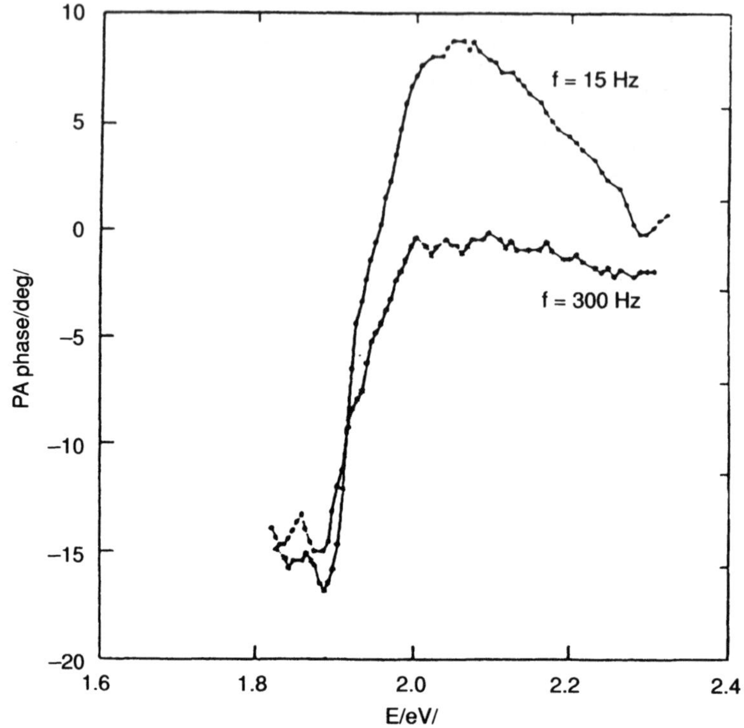

Figure 6.9(b) *Photoacoustic phase spectra of gallium selenide (Todorovic et al., 1991; reproduced with permission).*

6.6 Application 3 – dielectric materials

Rare earth metal oxides are often used to calibrate spectroscopic instruments, particularly in the visible region of the electromagnetic spectrum, because of their large number of sharp absorption peaks. Consequently, these materials have often been selected to test the capabilities of photothermal techniques for spectroscopic studies. The literature contains many spectra of these oxide materials obtained using PA, OBD, PTD, PPE and PTR detection. Figure 6.11 shows the photoacoustic amplitude spectrum of powdered samples of the rare earth oxides Nd_2O_3 and Ho_2O_3 across the spectral range 300–700 nm. A 1.5 kW Xe arc lamp was used as a light source. The observed peaks were found to correlate well with theoretical predictions.

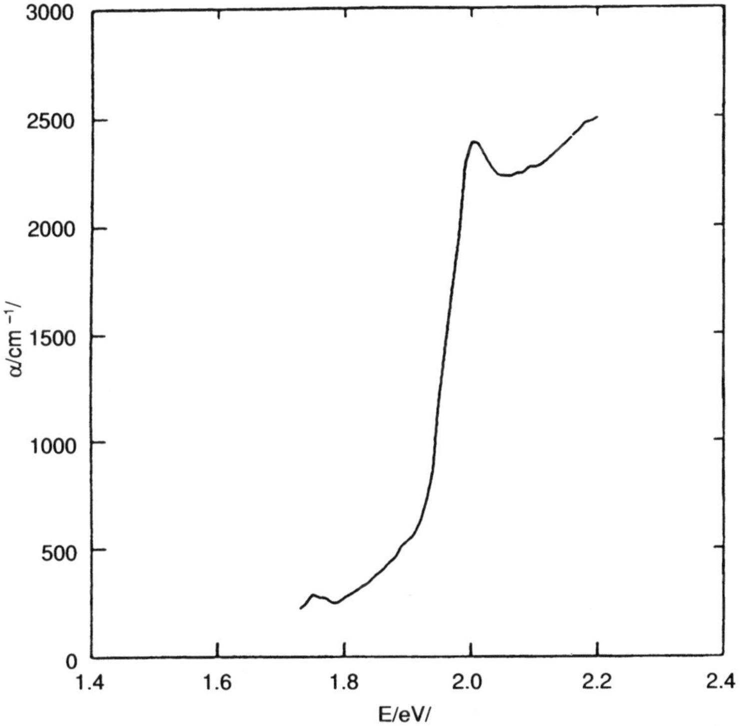

Figure 6.10 *The optical absorption coefficient spectrum of gallium selenide obtained from an analysis of the signal and phase data in Fig. 6.9 (Todorovic et al., 1991; reproduced with permission).*

The spectral depth profiling capabilities of photothermal methods, achieved by selection of the chopping frequency of the light source, are illustrated by the examples presented in Fig. 6.12. The test sample is a bi-layered material consisting of holmium oxide and neodymium oxide dispersed in a polystyrene matrix. At high modulation frequencies only the region close to the sample surface is contributing to the detected signal. A spectrum recorded at this frequency will only provide information on the surface layer. As the modulation frequency is reduced, an increasing contribution controlled by the spectral absorption characteristics of the second layer is found. The recorded spectrum at low frequencies shows characteristics of both sample oxides.

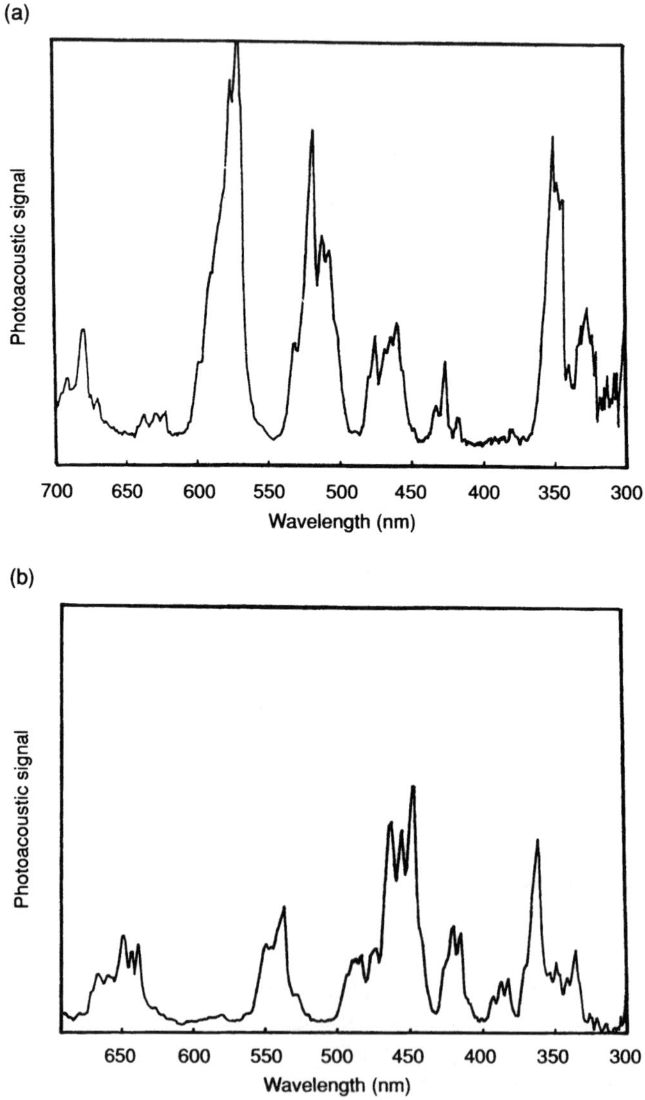

Figure 6.11 *Photoacoustic spectra of (a) neodymium oxide and (b) holmium oxide powder (Schoonover et al., 1984; reproduced with permission).*

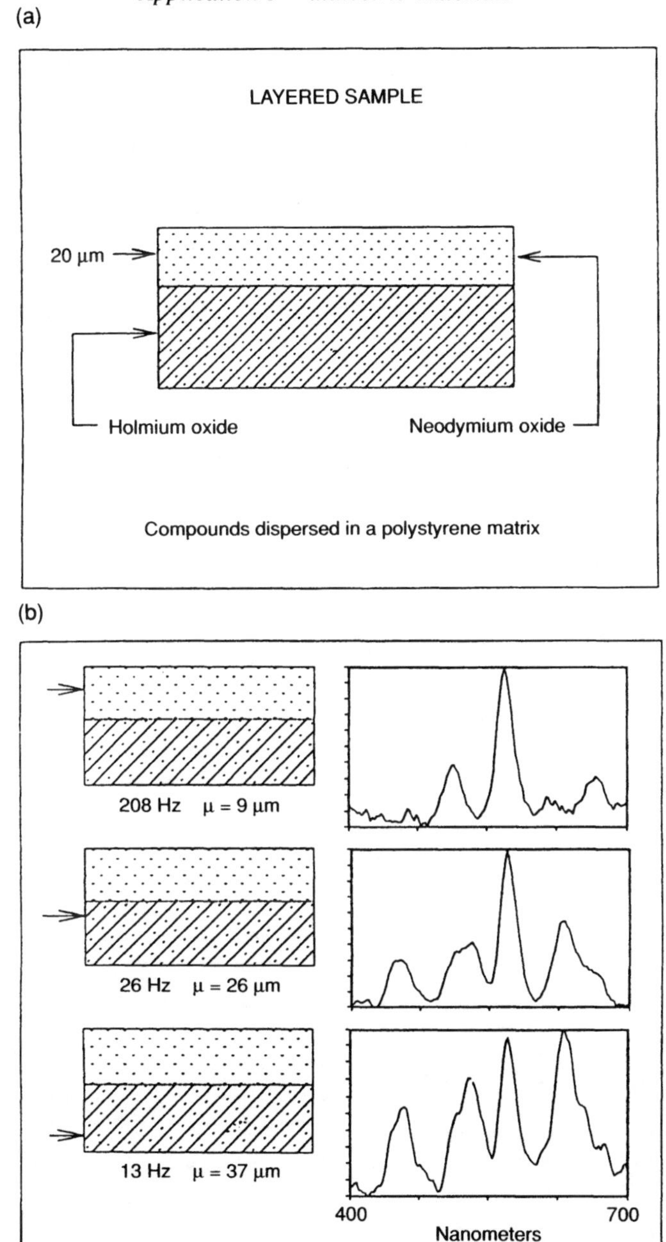

Figure 6.12 *(a) A layered sample of neodymium and holmium oxides and (b) the photoacoustic spectra obtained using three modulation frequencies to demonstrate the depth profiling effect (Wright et al., 1993; reproduced with permission).*

As an alternative to adjusting the modulation frequency for depth profiling, the phase reference angle may be varied such that the detected signal becomes weighted with near surface or deeper signal components. Figure 6.13 demonstrates this for the above bi-layered sample. The top curve is a spectrum recorded at a modulation frequency of 13 Hz. The middle curve is derived by maximizing the contribution from the surface component, and the lower curve by maximizing the signal contribution from the sample interior. The middle curve was obtained by adjusting the reference phase shift

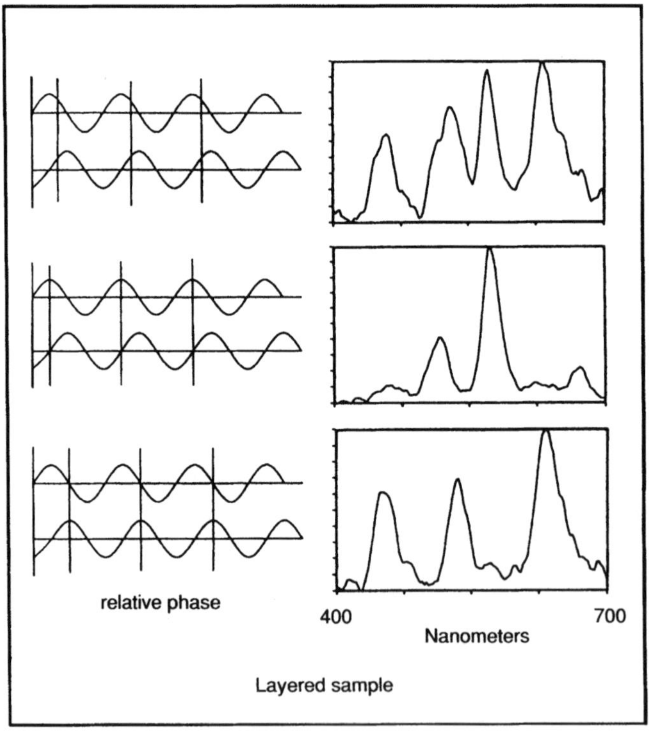

Figure 6.13 *A demonstration of depth profiling of the layered sample in Fig. 6.12 using phase analysis (Wright et al., 1993; reproduced with permission).*

so that a maximum signal component was obtained at a characteristic absorption peak for the surface layer. The bottom trace results from adjusting the reference phase angle to maximize the signal at the absorption peak due to the sample interior. A clear distinction between the upper and bottom layer spectral features can be seen.

6.7 Application 4 – organic materials

In the spectral analysis of polymeric materials the primary goal is to correlate absorption peaks to particular functional groups. The identification of functional groups in a given polymeric material provides a spectral fingerprint of the material. Their variation with chemical and thermal processing is commonly followed by IR spectroscopy. The use of photothermal spectroscopic methods for the investigation of polymeric materials is attractive because of the advantages listed in the introduction. PA, OBD, PZT and PPE methods, in combination with an FTS technique discussed in Section 6.1, have been used in studies of polymeric materials.

A set of nitrile plastic FTIR/PAS spectra are presented in Fig. 6.14 to demonstrate the ability of the technique to obtain good quality spectra from samples which have a range of different surface conditions.

With PA detection, it is important that the PA cell design is such that acoustic cell resonances are absent, minimized or removed to regions outside the bandwidth dictated by the mirror velocity and IR wavelength. Cell resonances can lead to distortions of the IR spectra. Another problem that may be faced is that of acoustic pick-up of environmental noises. The FTIR/OBD technique overcomes these problems associated with acoustic noise sources and offers high sensitivity. In addition, the absence of the requirement of a closed sample cell and the open geometry allows more flexibility in the size of samples used for spectroscopic analysis. The non-destructive nature of the technique is also important for some applications (for example in forensic science, archaeology and art history). Figure 6.15 shows a potential application of FTIR/OBD in the spectral characterization of ink on paper, demonstrating differences in the chemical composition of 'BiC' and 'Pilot' ink. Another example is presented in Fig. 6.16 of spectra of a fossil sample. The upper curve was obtained from a fossil insect embedded in a tar

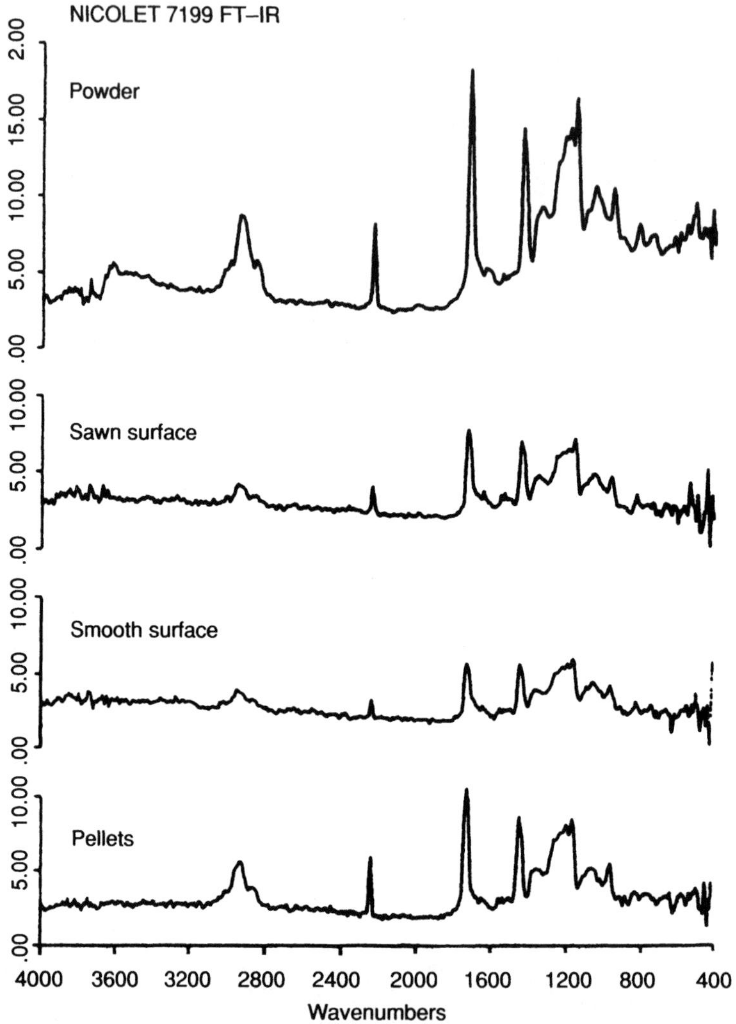

Figure 6.14 *FTIR photoacoustic spectra of nitrile plastic samples of the four types indicated in the figure (Vidrine, 1980, reproduced with permission).*

pit sample. The lower curves were from the tar matrix and an inner tar surface for comparison.

The use of a liquid medium to increase the OBD signal amplitude may pose problems in the IR region. The liquid medium must be transparent to both the IR heating radiation and the probe beam if heating and detection are to be on the same side of the sample

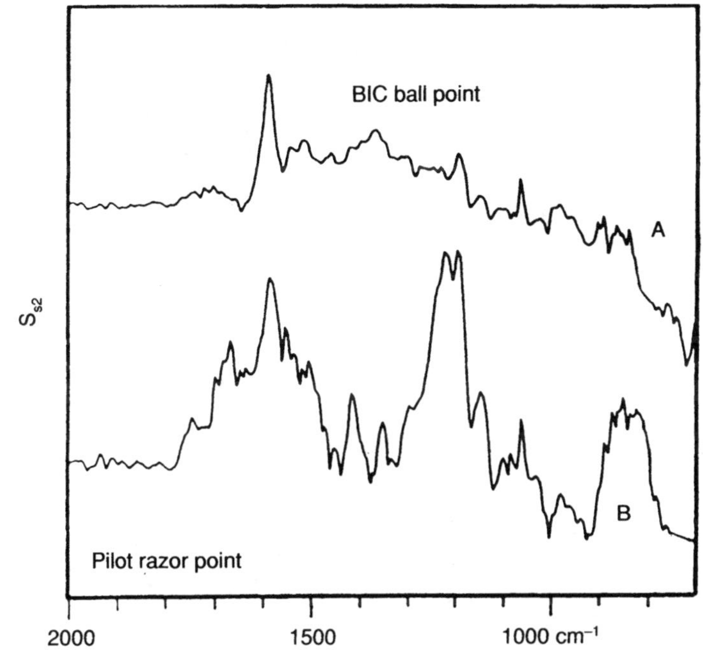

Figure 6.15 *FTIR/OBD spectra of two ink types obtained from samples of the inks on paper (Varlashkin and Low, 1986; reproduced with permission).*

surface. There are very few IR transmitting liquids covering the broad spectral range 400–4000 cm^{-1}, used in IR spectroscopy. Water has very strong IR absorption bands in this spectral region. CCl_4 has strong absorption below 1600 cm^{-1} but is transparent above. If the detection of the thermal wave is possible from below the sample, i.e. from its rear, then CCl_4 can be used over the whole spectral range as it is transparent to the HeNe probe beam generally used in mirage detection.

The PPE method avoids the difficulties of PA and OBD instrumentation. PPE-spectroscopy requires no closed cell and no coupling fluids (air or liquid). It is immune to acoustic noise and it eliminates spectral interference due to CO_2 and H_2O absorption because it can operate in a vacuum and because the sensor is usually in direct contact with the sample (via a grease layer). The sensor also offers a wide flat frequency response. PPE spectroscopy is ideally suited to spectral examination of materials deposited

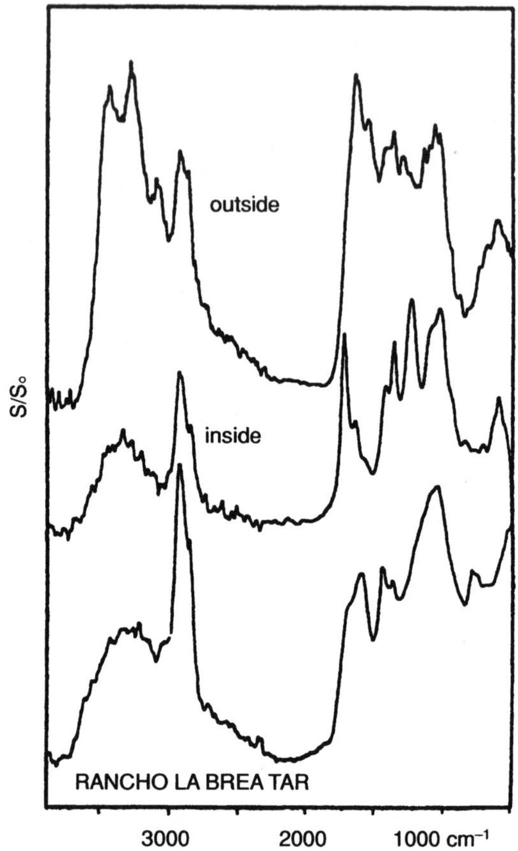

Figure 6.16 *Upper trace, OBD spectra of a fossil insect embedded in a tar pit sample; middle and lower traces are spectra obtained from the tar matrix and an inner tar surface (Low and Morterra, 1984; reproduced with permission).*

directly on the transducer film (which ensures good thermal contact with the sample). Signals measured from an optically thin film are directly proportional to the sample optical absorption coefficient.

6.8 Application 5 – biological and medical studies

Biological materials present conventional spectroscopic techniques with many difficulties owing to their inherent high light scattering properties and depth varying structures (blood, skin tissue, plants etc.). The abilities of photothermal techniques: to obtain spectra from such high light scattering media; to perform spectral depth profiling; to determine thermal properties and to study photochemical kinetics of biological materials have been demonstrated by various researchers.

6.8.1. Dermatology

The outermost part of human skin, the epidermis, provides a barrier which separates the interior from the environment. This barrier function is necessary to prevent excessive loss of water through the skin, as well as to prevent the entrance of noxious substances. There is great interest in making use of the skin as an entrance for properly designed drug formulations. The human epidermis is a tissue layer, roughly 100–500μm in thickness, composed of, from the inside to the outside, the stratum germinativum, the stratum granulosum, and the stratum corneum. The stratum corneum is responsible for the barrier function of the skin. It is a horny layer 6–20 μm thick, consisting of flattened keratinized cells.

Photoacoustic spectroscopy has the unique ability of providing depth-profiled analyses of the optical and thermal properties of skin.

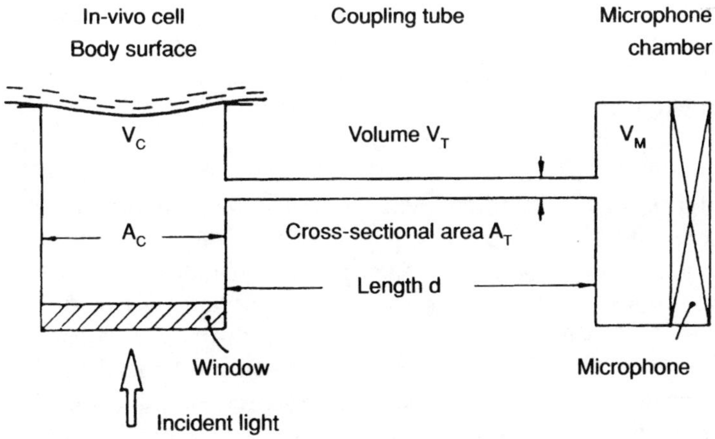

Figure 6.17 *Schematic diagram of a photoacoustic cell for use in the study of human skin (Giese et al., 1986; reproduced with permission).*

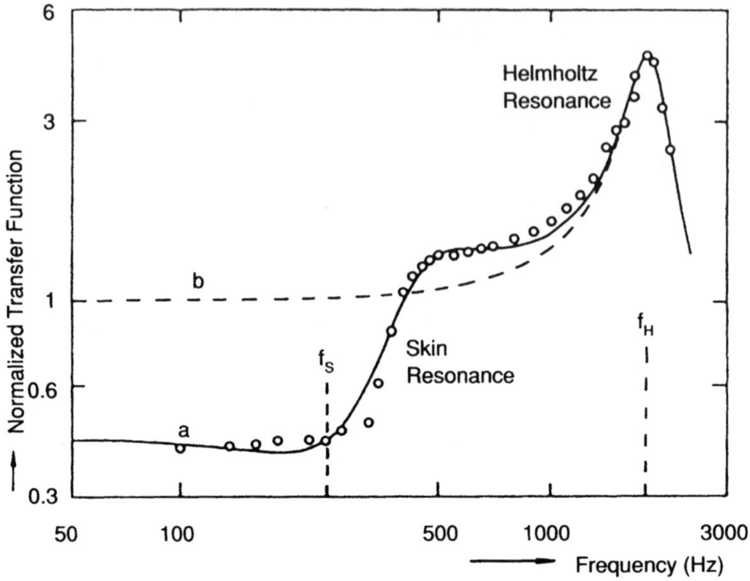

Figure 6.18 *Frequency response of the photoacoustic cell, Figure 6.17, coupled to human skin (Giese et al., 1986; reproduced with permission).*

In vivo studies on skin require special PA cells which allow the use of living skin and which avoid the noise generated by blood pulsations. Figure 6.17 is a schematic diagram of a PA cell used for such studies. Figure 6.18 shows the PA frequency response of human skin obtained in the UV–NIR spectral region. The thermal diffusion lengths in stratum corneum are about 10 and 4 μm at modulation frequencies of 180 and 1200 Hz.

The high absorption region in the UV results from the light absorption by protein (α-keratin) and in the NIR by water. The absorption of UV from sunlight by skin is particularly damaging, resulting in DNA strand breakage in the skin. The use of sunscreens to protect skin from UV damage has been studied photoacoustically. Figure 6.19 shows plots of UV absorption by two different sunscreens Eusolex 6300 and Eusolex 8020 *in vivo* at a chopping frequency of 1200 Hz. Another aspect which can be studied photoacoustically is the rate of penetration of sunscreen into skin. During the diffusion of a sunscreen into skin, the levels of pigments in a surficial layer decrease, which leads to a loss of PA signal from this layer. From the time dependent variation of the PA signal, see

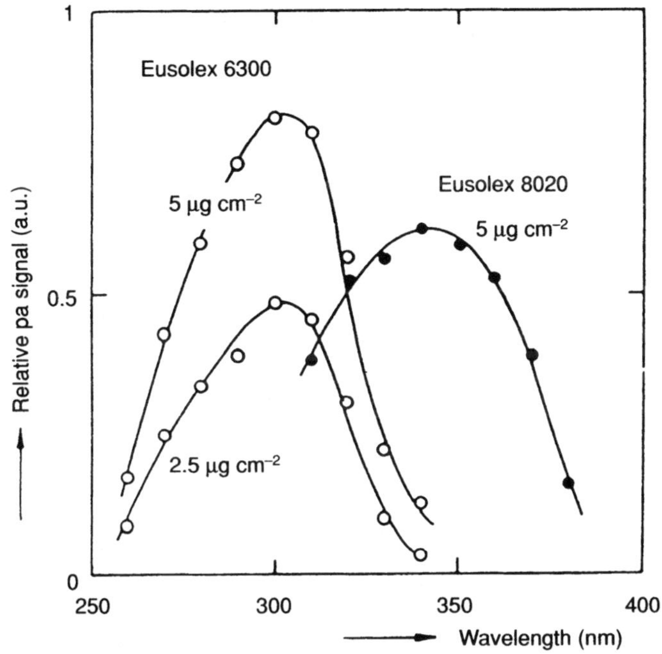

Figure 6.19 *Photoacoustic measurements of the UV absorptions of the sunscreens Eusolex 6300 and 8020 applied to human skin (Giese et al., 1986; reproduced with permission).*

Fig. 6.20, it may be possible to determine sunscreen penetration rate and its time of residence in different skin layers. These parameters are important in characterizing the usefulness of these topical creams for the efficient protection of the skin against UV radiation.

6.8.2 Hematology

Hemoglobin, the protein responsible for the O_2–CO_2 exchanges between blood and tissue, has a tetraprophyrin cycle bound to its amino acid skeleton, which gives it a reddish colour. The advantages of PAS in this field of research are:

1. its ability to detect accurately some compounds present in highly diffusive media such as blood and in non-uniform and opaque materials, such as the surfaces of organs;

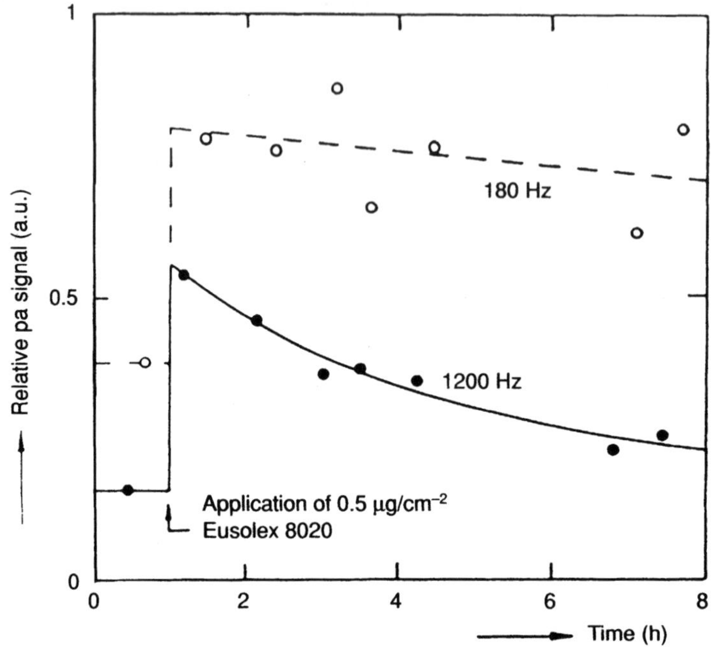

Figure 6.20 *Time dependence of photoacoustic signals from Eusolex 8020 (Giese et al., 1986; reproduced with permission).*

2. the avoidance of any complicated chemical extraction and purification processes, limiting manipulations;
3. that it requires only a small amount of the substance and is non-destructive, permitting further investigations by other techniques.

A typical PA spectrum of hemoglobin is shown in Fig. 6.21 with the three major bands at 415, 540 and 580 nm. Spectra of blood from patients having different illnesses are also shown in this figure. The changes from normal are easily identified.

The photoacoustic technique can also be used to follow the sedimentation process in blood. At a fixed modulation frequency the sample region contributing to the PA signal is fixed. Any changes in the amount of chromopores in this region should induce a change in the PA signal. In a blood sample, sedimentation induces movement of erythrocytes (red cells) to the bottom. This decreases the

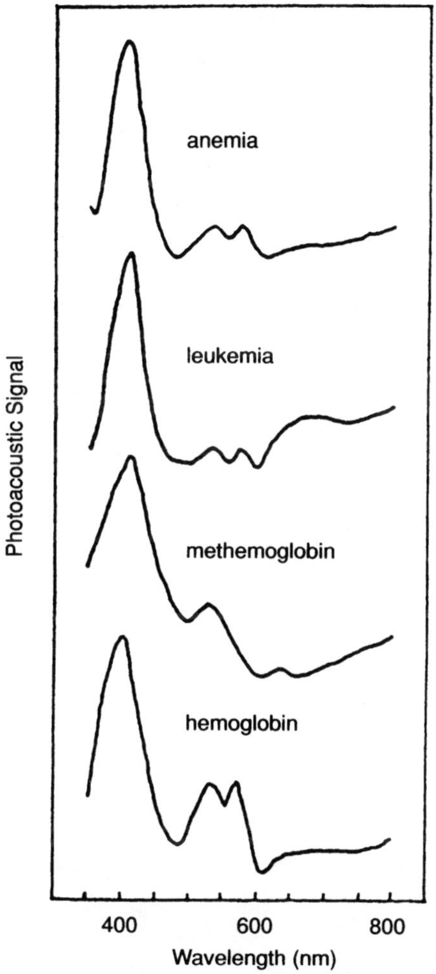

Figure 6.21 *Photoacoustic spectra of hemoglobin, methemoglobin and of the blood of patients suffering from the illnesses indicated in the figure (Qing Pan et al., 1987; reproduced with permission).*

amount of pigment in the upper layers of the sample. As a result both PA amplitude and phase change. Following this process in time allows the determination of the sedimentation rate.

6.8.3 Photosynthesis

Photoacoustic techniques have been used extensively for studies of the effects of environmental stresses on photosynthetic activity in plants. The absorption of visible radiation by plants and the eventual release of this energy can be expressed as:

$$E_{abs} = E_{heat} + E_{luminescence} + E_{photochemistry}$$

The energy lost by luminescence processes is mostly due to fluorescence (emission of a photon of lesser energy than the one which was absorbed) and accounts in a leaf for up to 3% to 5% of all deactivations. This is a small amount and can be neglected from the above equation. Thus, the energy transformed into heat can be expressed as:

$$E_{heat} = E_{abs} - E_{photochemistry}$$

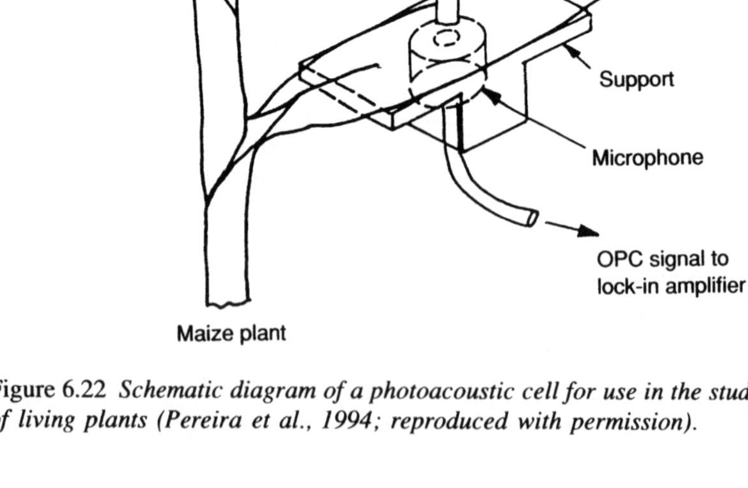

Figure 6.22 *Schematic diagram of a photoacoustic cell for use in the study of living plants (Pereira et al., 1994; reproduced with permission).*

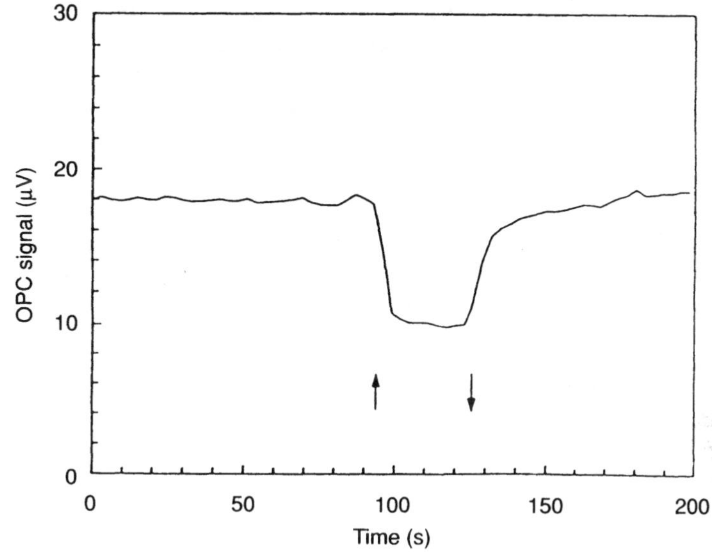

Figure 6.23 *Change in photoacoustic signal from a plant leaf whilst a white light source is switched on ↑ and off ↓ (Pereira et al., 1994; reproduced with permission).*

The energy lost via a photochemical mechanism results in energy storage by chemical intermediates and in oxygen evolution. Photoacoustics can be used to study these absorption and photochemical changes.

Figure 6.22 is a schematic diagram of an *in vivo* PA apparatus which has been developed to monitor oxygen evolution. Figure 6.23 shows the signal change occurring as a dc white light source was switched on ↑ and then off ↓. Whilst the dc light was off the modulated light produced a significant PA signal component by a photobaric process. Modulated light produces a modulated oxygen evolution by photosynthesis. This flows into the PA cell and produces a modulated pressure component that results in a microphone signal. When a sufficiently high dc optical intensity is applied the photosynthesis process is saturated and the response to the additional periodic illumination is lost – resulting in the dip in signal shown in Fig. 6.23. The magnitude of the signal drop can be calibrated to provide a measure of oxygen evolution rate. This PA technique has the potential of providing a compact means of monitoring oxygen evolution by living plants.

Photoacoustic techniques have also been used to study detrimental effects of water stress, high and low temperatures, high and low light fluxes, and airborne pollutants and herbicides on leaves and entire algal cells.

References

Futamata, M., Fuji, A., Atsuya, I. and Sata, C., 'Prototype instrument for PAS and examples of PA characteristics of solid specimens', *Trans. Jap. Welding Soc.*, **23** (1992) 3–8.

Giese. K., Nicolaus, A., Sennhenn, B. and Kolmel, K., 'Photoacoustic *in vivo* study of the penetration of sunscreen into human skin', *Can. J. Phys.*, **64** (1986) 1139–41.

Low, M.J.D, and Morterra, C., 'Infrared spectra of fossils', *Appl. Spectrosc.*, **38** (1984) 807–12.

Pereira, A.C., Neto, G.O., Varagas, H., Cella, N. and Miranda, L.C.M., 'On the use of the open photoacoustic cell technique for studying photosynthetic O_2 evolution of undetached leaves: Comparison with Clark-type O_2 electrode', *Rev. Sci. Instrum.*, **65** (1994) 1512–16.

Pan, Q., Qin, S., Zhang, S., Zhang, J. and Zhu, S., 'Applications of photoacoustic spectroscopy to human blood', in *Springer Series in Optical Sciences Vol. 58: Photoacoustic and photothermal phenomena* (eds P. Hess and J. Pelzl) (1987), 542–5.

Rosencwaig, A., 'Photoacoustic spectroscopy: a new tool for investigating solids', *Anal. Chem.*, **47** (1975) 592A.

Schoonover, J.R., Lee, Y.L., Su, S.N., Lin, S.H. and Eyring, L., 'Photoacoustic spectroscopy of rare earth oxides', *Appl. Spectrosc.*, **38** (1984) 154–8.

Todorovic, D.M., Nikolic, P.M. and Ristovski, Z.D., 'Quantitative determination of absorption using coupled amplitude and phase photoacoustic spectra', in *Physical acoustics* (eds O. Leroy and M.A. Breazeale), Plenum Press, New York (1991) 641–6.

Tokumoto, H., Tokumoto, M. and Ishiguro, T., 'Photoacoustic spectra of semiconductors in the strong absorption region', *J. Phys. Soc. Jpn.*, **50** (1981) 602–5.

Varlashkin, P.G. and Low, M.J.D., 'FT–IR photothermal beam deflection spectroscopy of black inks on paper', *Appl. Spectrosc.*, **40** (1986) 507–13.

Vidrine, D.W., 'Photoacoustic Fourier transform infrared spectroscopy of solid samples', *Appl. Spectrosc.*, **34** (1980) 314–19.

Wright, S.L., Hammaker, R.M. and Fateley. W.G., 'Hadamard transform photoacoustic spectometry and depth profiling', *Appl. Spectrosc.*, **47** (1993) 272–75.

7
Non-destructive evaluation

Non-destructive evaluation (NDE) encompasses a range of techniques that have been developed to evaluate the integrity and/or quality of engineering components. NDE is dominated by the problem of detecting, imaging and sizing structural defects, such as cracks, which might lead to a component failure. It also includes non-destructive methods that provide information about the characteristics of the materials utilized in the manufacture of components. NDE has a well established vital role in maintaining the safety of air travel and a very important part to play in ensuring the safe and reliable operation of major power generation and chemical plants.

The most widely used NDE techniques are: ultrasonics; X-ray radiography; eddy current, dye-penetrant and visual inspection. Although thermal techniques have yet to make a major impact in the NDE field, they are now being used to solve an increasing number of industrial inspection problems and they are being investigated enthusiastically in laboratories throughout the world. They have significant advantages over the traditional NDE methods in a number of new important areas, notably for the inspection of the new materials, ceramics, composites and coatings, that are becoming essential to modern industry.

It must be recognized at the outset that thermal/thermal wave techniques are intrinsically surface biased. They are techniques that offer an attractive alternative to ultrasonics for sub-surface, near-surface inspection but they will never be a sensible alternative for deep probing to locate hidden cracks. The essential characteristics of thermal/thermal wave techniques are:

1. an ability to detect interfaces between media of differing thermal properties;

2. well developed non-contact generation and detection techniques;
3. the potential for rapid large area scanning for defects.

The surface-specific nature of the techniques has led to a great deal of interest in coatings inspection and the testing of laminated composite materials. A very wide range of coatings and laminated materials are used by industry and none of the traditional NDE techniques are particularly well suited to their inspection.

7.1 Defects in engineering components

Fig. 7.1 is a schematic diagram of some of the main structural imperfections that may be present in a coating on an engineering component. A poor bond between the coating and the substrate may lead to the failure of such a component. The bond may be partially or completely absent (i.e. a delamination). The coating structure may vary because of local sites of high porosity or the presence of foreign particles. The coating thickness may vary and its surface may be rough. Internal stresses in the coating layer may be produced during the deposition process leading to the development of cracks

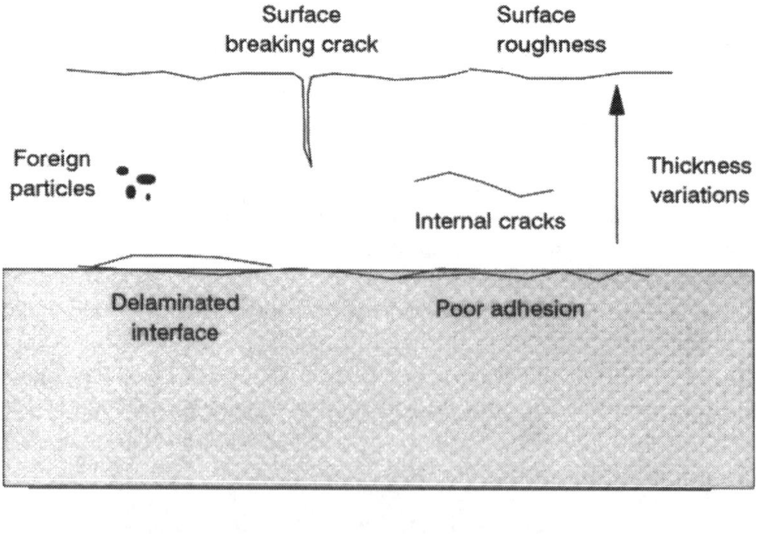

Figure 7.1 *Schematic diagram of defects in a coating.*

both in the plane parallel to the coating surface and perpendicular to it. Some similar defects, at or near the surface, may be found in composite material components. Delaminations between layers can occur in processing or develop because of internal stress; regions of porosity or poor fibre adhesion can occur in the matrix and cracks may develop because of overloading or other types of damage. And in engineering ceramics, strength is determined by minute cracks or voids at the surface which cause stress concentrations leading to brittle fracture. For each of these important groups of materials there are good reasons to consider the surface biased thermal NDE techniques that are discussed in this chapter.

7.2 Frequency-domain, thermal wave NDE

The potential of thermal waves to become a new NDE probe has stimulated extensive investigations in numerous laboratories throughout the world. Coating thickness assessment, which is a particularly favorable application for thermal waves, will be covered first in this section before dealing with the main NDE activity of defect detection. Much of the theoretical background for thermal wave NDE will be found in Chapter 2.

7.2.1 Coating thickness determination

Coating thickness measurements are required for almost all industrial applications of surface coatings. Thickness measurements are necessary for quality assurance during component manufacture and may subsequently be necessary as a part of in-service inspection. This is particularly true of coatings used to provide wear resistance for which thickness values are needed to provide an assessment of residual life.

Most of the photothermal methods (PA, OBD, PTR, MOR, PZ) have been shown to be capable of detecting thickness variations in a variety of coating/substrate systems. Measurements of thermal wave phase, in preference to amplitude, are usually used in coating thickness assessment as these are unaffected by variations in the optical reflectivity of the sample. (The thermal wave amplitude shows strong variations with the reflectivity of a surface.) Thermal wave interference causes a systematic variation of thermal wave phase with coating thickness, as discussed in Chapter 2. Figure 7.2

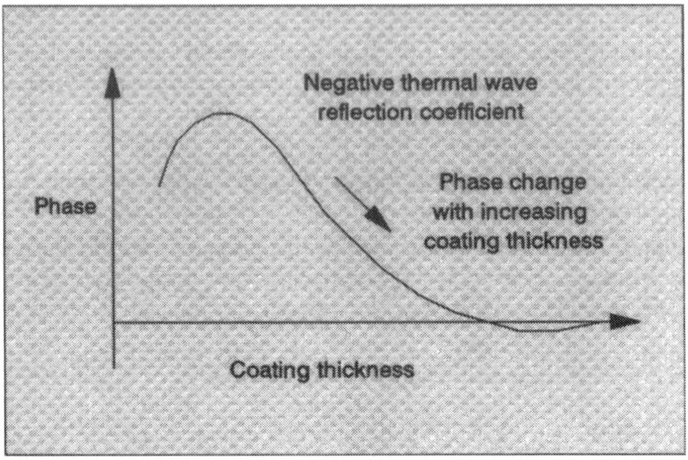

Figure 7.2 *Schematic diagram of the variation of phase with coating thickness.*

shows the general form of this variation of phase with coating thickness, for an optically opaque coating on a thermally thick substrate. An example of some experimental PTR phase angle data obtained from a wedge of plasma sprayed molybdenum coating on a steel substrate is shown in Fig. 7.3. Several features that are characteristic of the thermal wave interference phenomenon can be seen in this figure. For small thicknesses the phase angle rises with thickness to a peak value. With further increases in the coating thickness the phase angle decreases. This behavior is characteristic of a coating–substrate combination that results in a negative value of the thermal wave reflection coefficient (obtained where the effusivity of the coating is less than that of the substrate). The variation with the modulation frequency of the thermal wave used to probe the coated sample demonstrates the depth profiling capability that is an important characteristic of photothermal techniques. With decreasing frequency, the peak occurs at increasing coating thicknesses, extending the range of thicknesses that can be probed. In practice, the modulation frequency is chosen to provide the best coverage of the range of coating thicknesses to be measured. It is usual to adjust the frequency so that the peak of the phase–thickness characteristic occurs at a coating thickness below that to be measured. A single valued, comparatively linear part of the

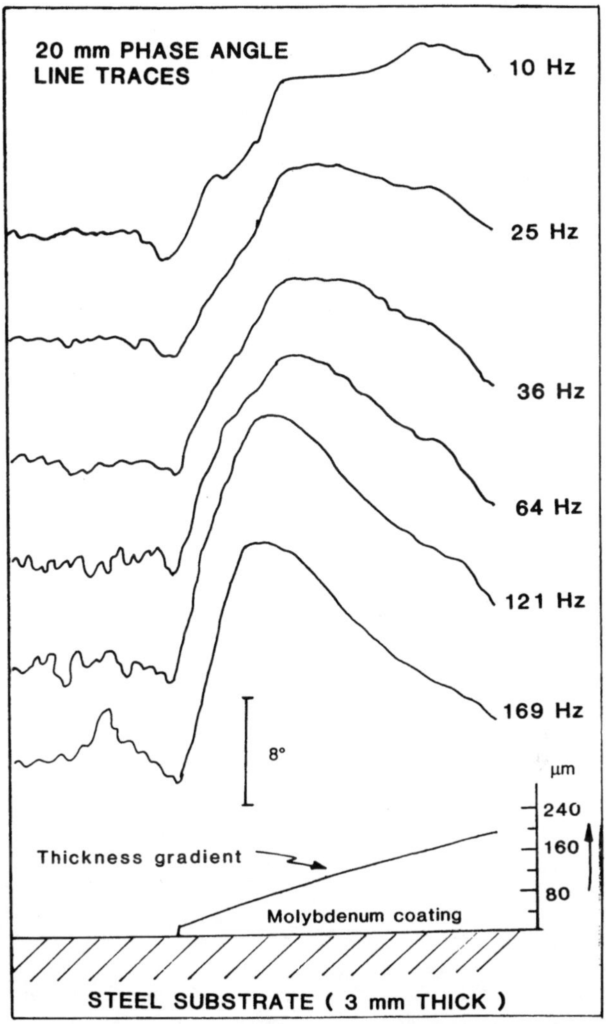

Figure 7.3 *A set of PTR phase angle line traces over a thickness gradient of plasma sprayed molybdenum coating on mild steel at the modulation frequencies indicated.*

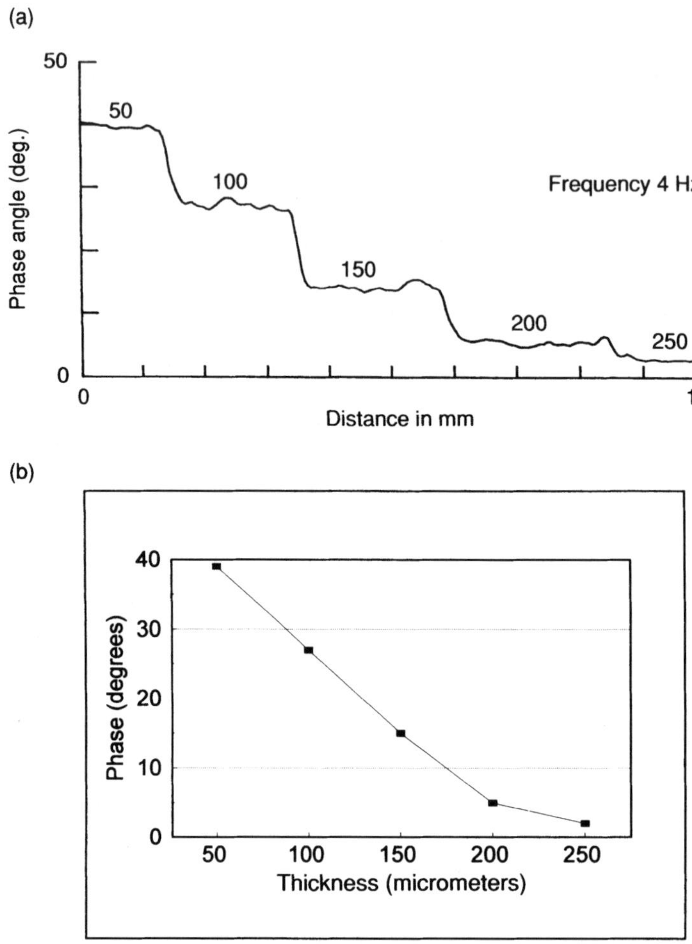

Figure 7.4 *(a) A PTR phase angle line trace along an yttria stabilized zirconia coated test strip with coating thickness steps of 50, 100, 150, 200 and 250 μm. (b) A phase thickness calibration obtained from the data in (a).*

phase–thickness characteristic can then be used for thickness measurement, after calibration against known coating thickness samples (see below).

Thermal wave thickness gauging is not restricted to opaque coatings. Thickness variations in translucent coatings can also be detected as is illustrated in Fig. 7.4. Figure 7.4(a) shows the phase angle variation along a steel sample coated with a number of different thicknesses of the translucent thermal barrier coating zirconia. Each of the step changes in coating thickness on the test piece produced a distinct step change in the phase of the thermal wave. No peak in phase was observed in this case because even on the thinnest coating step, 50 μm, the ratio of coating thickness to thermal diffusion length, L/μ, exceeded the value of about 0.3 at which the thermal wave interference peak occurs, see Fig. 2.8. The variation in phase obtained on each step was caused by variations in the actual thickness of the coating. The data obtained from a stepped coating thickness test piece of this type provides the phase thickness calibration needed for thickness gauging. The calibration obtained from the data in Fig. 7.4(a) is shown in Fig. 7.4(b).

In thickness gauging, translucency can be beneficial because a greater range of coating thicknesses can be examined at a particular modulation frequency. This is because of the reduced attenuation of thermal waves arriving at the sample surface generated by heat sources at a range of depths within the coating (as discussed in Section 3.3).

Thermal interferometry provides an attractive means of coating thickness measurement that has been found to be suitable for a wide range of coating types. The sensitivity of the technique is determined by the magnitude of the coating/substrate thermal wave reflection coefficient and the signal-to-noise ratio of the detection system. It is common to achieve a sensitivity amounting to 1% or 2% of coating thickness and to be limited in absolute accuracy by uncertainties in the true thicknesses of samples used as calibration standards. The particular advantages of this thermal wave technique over other means of coating thickness measurement are that it is:

1. non-contacting;
2. rapid;
3. readily calibrated and simple to interpret;
4. suitable for computer control and data logging;
5. applicable to the measurement of a wide range of coatings which may have rough unprepared surfaces.

7.2.2 Sub-surface defect characterization

Sub-surface defects such as air gaps and partially contacting internal surfaces can be detected by photothermal techniques because

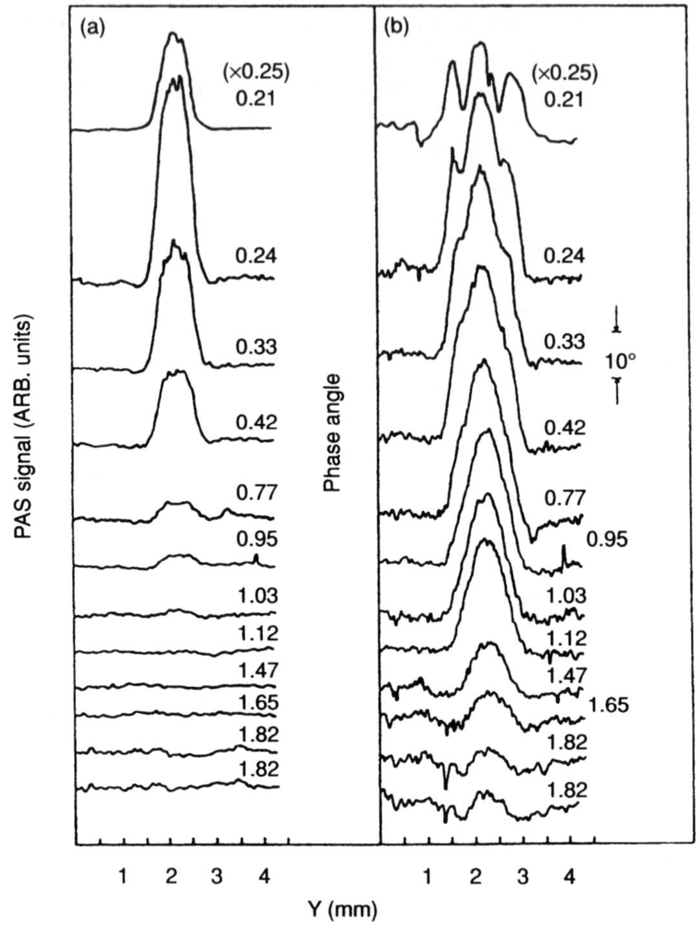

Figure 7.5 *Measurements of (a) the amplitude and (b) the phase of the PA signal over a channel cut beneath the surface of an aluminum sample. The values of the ratio of aluminum thickness to thermal diffusion length at the measurement frequency of 600 Hz are indicated for each trace (Thomas et al., 1980; reproduced with permission).*

both of these defect types present an impedance to heat flow. The detection of partially contacting surfaces is perhaps the more important as this defect type is a precursor to the discrete airgap crack type defect. The application of thermal wave techniques to defect detection and characterization began soon after the revival of the PA effect in solid materials. Thomas *et al.* (1980) provided one of the first demonstrations of the NDE potential of photothermal techniques using the PA effect for the detection of a sub-surface channel machined in an aluminum test piece. The channel was cut at an angle to the surface to provide a sub-surface interface at a range of depths to facilitate an investigation of the influence of defect depth on the response of the technique. The results obtained, Fig. 7.5, highlighted several basic features of thermal wave inspection:

1. significant changes in both signal amplitude and phase angle over the channel (simulated sub-surface defect);
2. a greater sub-surface structure detection sensitivity in the phase angle than in the signal amplitude measurements;
3. the potential for depth profiling through variation of the thermal wave frequency.

The overall shapes of the defect signal and phase traces depend on the defect geometry and on several experimental parameters such as the modulation frequency, the spatial extent of the optical heat source and detection localization. When heat flow can be regarded as being uni-dimensional, simple analytical relationships can be used to model the photothermal signal variations from a planar sub-surface defect (airgap or contact resistance). These two defect models were introduced earlier in Section 2.6, and the relevant formulae are shown in Table 7.1.

Table 7.1 Complex surface temperature

Opaque coating	$T_s = \frac{(1-R(\lambda))I_0}{2k_c\sigma_c}\left[\frac{1+\Gamma\exp(-2\sigma_c L_c)}{1-\Gamma\exp(-2\sigma_c L_c)}\right]e_{j\omega t}$
Air gap	$\Gamma = \frac{\Gamma_1+\Gamma_2\exp(-2\sigma_g L_g)}{1+\Gamma_1\Gamma_2\exp(-2\sigma_g L_g)}$
	$\Gamma_1 = \frac{k_c\sigma_c - k_g\sigma_g}{k_c\sigma_c + k_g\sigma_g} \quad \Gamma_2 = \frac{k_g\sigma_g - k_s\sigma_s}{k_g\sigma_g + k_s\sigma_s}$
Thermal contact resistance	$\Gamma = \frac{1-b_1+R_T k_s\sigma_s}{1+b_1+R_T k_s\sigma_s} \quad b_1 = \frac{k_s\sigma_s}{k_c\sigma_c}$

Indices c, g and s refer to the coating, air-defect and substrate layer respectively.

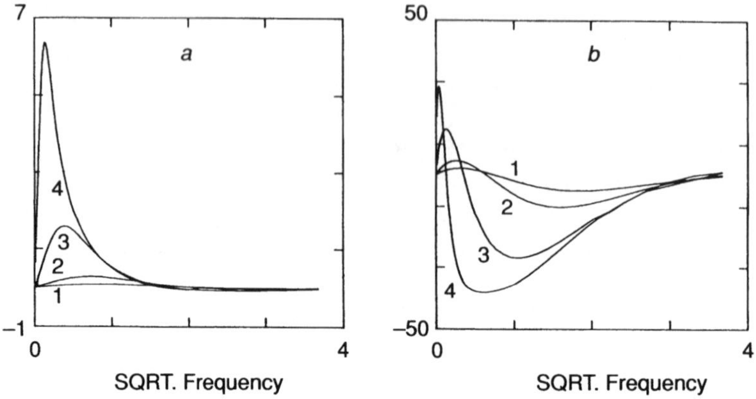

Figure 7.6 *Graphs of the theoretical (a) surface temperature and (b) phase dependence on frequency for air-gap defects of: (1) 0.1; (2) 1.0: (3) 10.0 and (4) 100 μm thickness in stainless steel at a depth of 0.5 mm.*

Figure 7.6 shows the theoretical frequency dependences of the normalized surface temperature and its phase caused by a range of different thickness airgap defects lying in stainless steel at a depth of 0.5 mm. The normalization is with respect to the complex temperature that would be obtained in a semi-infinite sample without a defect. At low frequencies and for large defect thicknesses a strong enhancement in the surface temperature is predicted. With increasing frequency this temperature enhancement reduces and vanishes as the sample becomes thermally thick.

The phase angle dependence, Fig. 7.6(b), at low modulation frequencies and large airgap thicknesses becomes positive, i.e. is in advance. With increasing frequency this advance turns into a phase lag. With further increases in the frequency the phase lag decreases and eventually diminishes to zero. This characteristic phase variation from a sub-surface defect has been observed experimentally in homogeneous and coated samples. A clear example of this behavior is shown in Fig. 7.7, obtained using PTR from an aluminum coated steel sample which had a 3 mm wide adhesion defect at the coating–substrate interface.

An expansion of the surface temperature expressions in the low frequency limit results in the following equations for the two defect models:

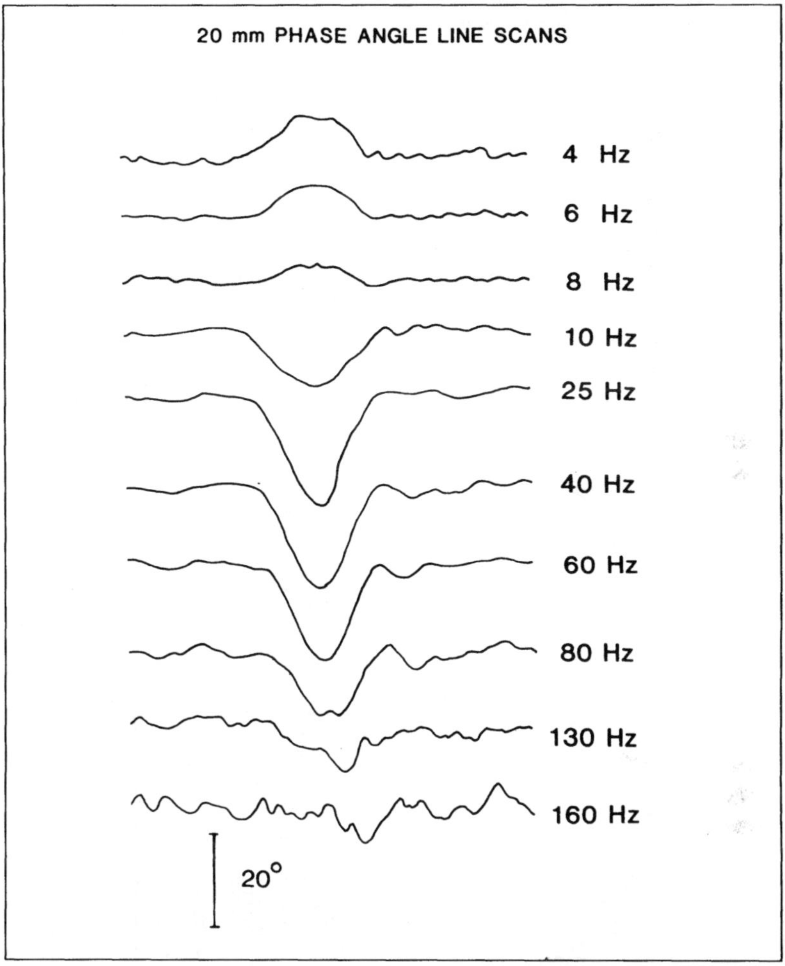

Figure 7.7 *Traces of PTR phase across a 3 mm defect at the interface between a sprayed aluminum coating and a steel substrate.*

Air-gap
$$T_s \approx T_0\left(1+\left(\frac{k_s\sigma_s}{k_g\sigma_g}\right)\sigma_g L_g\right) \tag{7.1}$$

Contact resistance
$$T_s \approx T_0(1+R_T k_s \sigma_s) \tag{7.2}$$

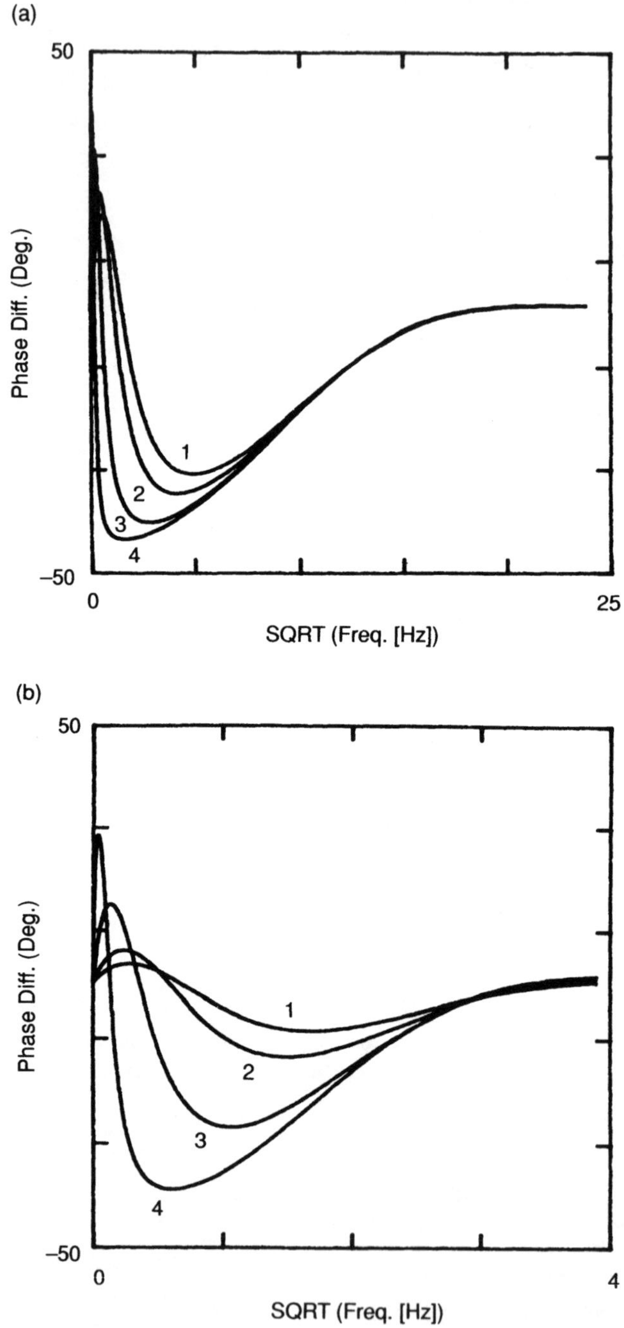

Figure 7.8(a) and (b).

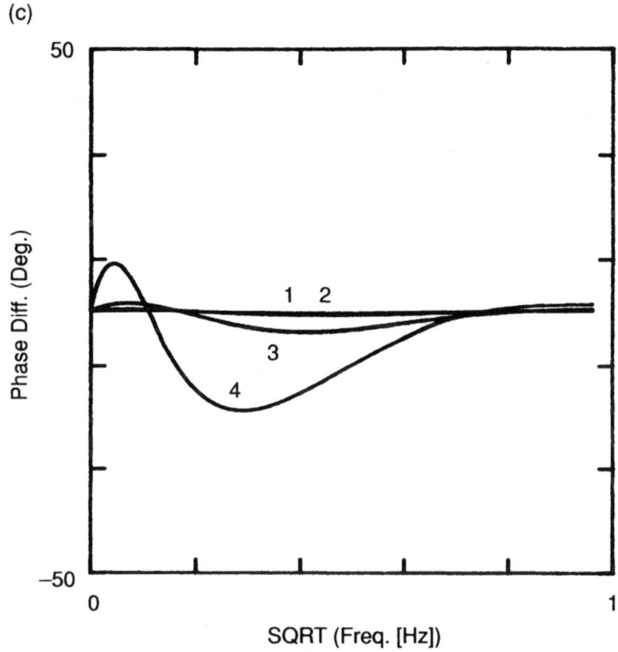

Figure 7.9 *Graphs of the theoretical phase dependence on frequency for air-gap defects of: (1) 0.1; (2) 1.0: (3) 10.0 and (4) 100 μm thickness in: (a) aluminum; (b) stainless steel and (c) epoxy resin at a depth of 0.5 mm.*

Both equations correctly predict the phase advance. In addition these expressions emphasize another important characteristic of subsurface defect detection: the influence of the host material. For the air-gap case this appears as the thermal wave impedance ratio of the host material to the air enclosed by the defect: $k_s\sigma_s/k_g\sigma_g$. The larger the value of this term the greater will be the phase advance and the overall defect phase variation for a given airgap thickness. Figure 7.8 shows the theoretical phase variations caused by a set of different thickness air-gap defects 0.5 mm below the sample surface in three materials aluminum, stainless steel and epoxy resin, whose thermal impedance ratios to air are 24 000:1; 7 200:1; 550:1 respectively. The figures show that the magnitude of the effect of a defect on thermal wave phase is highly dependent on the thermal properties of the host material. Air-gaps of less than 10 μm produce large effects in an aluminum component but such small effects in an epoxy component that they would be undetectable. For an inter-

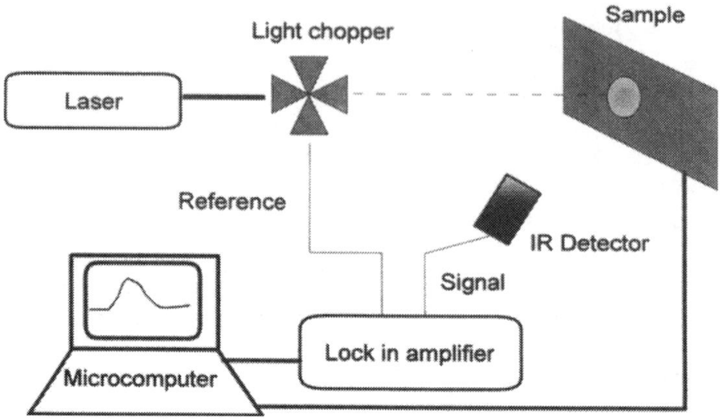

Figure 7.9 *Block diagram of a PTR inspection system.*

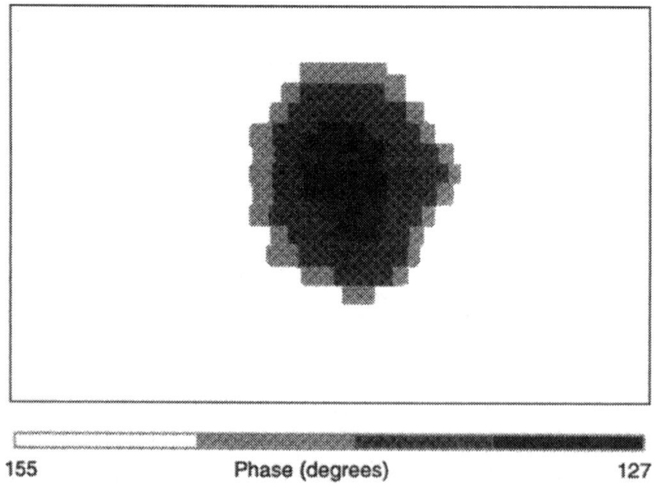

Figure 7.10 *A PTR thermal wave phase image of a 10 mm diameter air gap defect lying 1 mm beneath the surface in a mild steel block.*

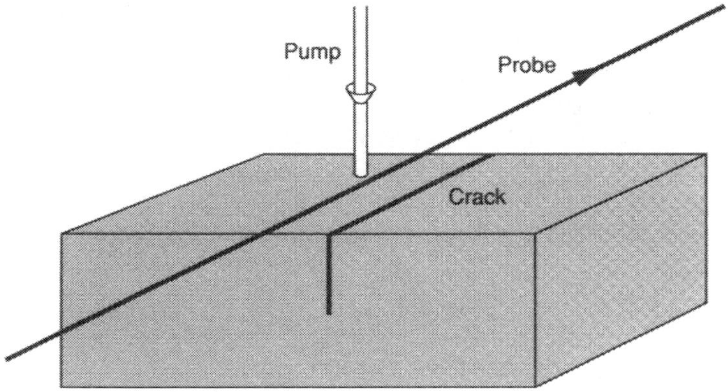

Figure 7.11 *Schematic diagram of an OBD system arrangement employed to detect vertical cracks.*

face producing a thermal contact resistance, the defect enhancement factor is dependent directly on the thermal effusivity of the host material, equation (7.2).

Sub-surface defect detection has been demonstrated using systems based on the OBD, PTR , PA and PZ techniques. A block diagram of a PTR NDE system is shown in Fig. 7.9. Images of the part being tested are produced by scanning the component through the laser beam in a raster, c-scan, fashion and recording at each point the thermal wave signal and phase. Figure 7.10 is an example of a phase image of a sample, indicating the position and size of a sub-surface defect. In practice it is always necessary to experiment with modulation frequency to maximize defect detection sensitivity. This amounts to finding the thermal diffusion length that corresponds to the defect depth to maximize thermal wave interference in the space between the defect and the surface.

Vertical surface breaking cracks can also be detected photothermally. An asymmetric probe detection scheme is usually employed for this type of defect. The OBD technique has been shown to be effective using the transverse probe beam deflection to detect a vertical crack. In the test arrangement, shown in Fig. 7.11, the probe beam is positioned ahead of the pump beam. As the beams approach the crack an enhancement of deflection amplitude is produced as lateral heat flow through the crack is impeded. When the probe beam is over the crack the signal reduces

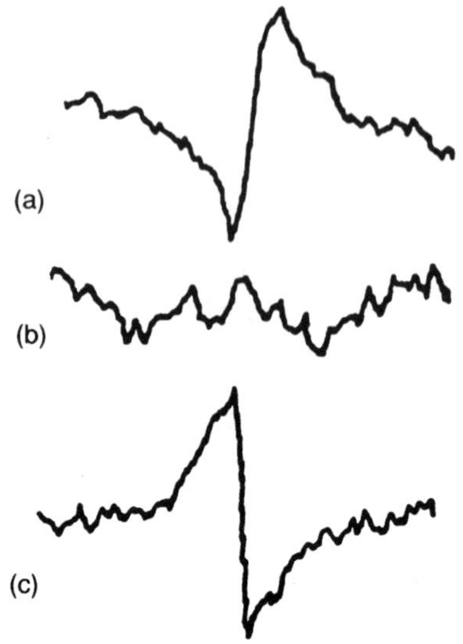

Figure 7.12 *Experimental data obtained using the OBD (mirage) technique from a sample containing a vertical surface breaking crack where (a) the heating beam is to the left of the probe beam, (b) the beams intersect and (c) the heating beam is to the right of the probe beam (Grice et al., 1983; reproduced with permission).*

to zero and as it moves further away from the crack plane the deflection amplitude drops further due to the restriction of heat flow from the pump beam side of the crack plane. The phase variation over the crack undergoes a 180° phase change. This signal variation is characteristic of a vertical crack. Figure 7.12 shows examples of experimental data obtained at a vertical crack using the OBD technique.

Detection of vertical cracks by PTR has also been achieved using an asymmetric detection scheme comprising an infrared detector focused on a spot in advance of the pump beam. Signal variations similar to those observed by OBD are found. The signal amplitude also shows strong contrast due to the enhancement of the surface emissivity by the partially open crack.

7.2.3 Limits of resolution – thermal wave microscopy

As for all NDE systems, it is essential to understand the factors that affect the resolution of a photothermal NDE system. The parameters which determine the spatial and depth resolution are:

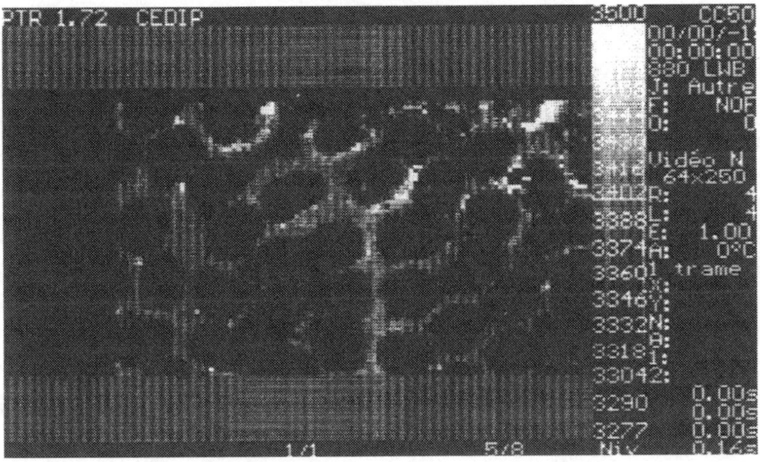

Figure 7.13 *MOR phase image of carbon fiber bundles in a 3-D C/C composite. The scanned area is 50 μm × 30 μm and a step size of 0.5 μm was employed. The phase scale in 0.1° units is at the right of the image (F. Lepoutrere, ONERA, Chatillon, France; reproduced with permission).*

1. the spot sizes of both the source (pump) beam and the probe;
2. the depth and geometry of the sub-surface structure;
3. the modulation frequency.

The resolution of a thermal wave imaging system is **not** determined by the familiar Rayleigh criterion. Unlike most other systems, such as optical, electron optical and ultrasonics, thermal wave images are formed in the extreme near field limit rather than at the Fraunhofer limit. Because of the decay of thermal waves, the object plane (containing a defect) must be within about a thermal diffusion length of the surface image plane, since the thermal diffusion length = (thermal wavelength)/2π, object and image planes are separated by a fraction of a wavelength – the extreme near field limit. In this limit the image is effectively a contact print of the object and is independent of wavelength. It has been shown that small, 30 μm, artificial sub-surface structures can be imaged by thermal waves with wavelengths of many millimeters. The resolution of nearby sub-surface structures is strongly dependent on the pump beam and probe spot sizes. If the optical spot size is much larger than the thermal diffusion length, lateral resolution will be determined by the optical spot size. The depth resolution will still be set by the thermal diffusion length. Besides the above parameters, resolution can also be limited by engineering details of the inspection system such as the step sizes of the sample–optical beam scanning system and focusing lens imperfections.

Numerous research groups have built thermal wave microscopes and obtained high resolution images of thermal microstructure. Figure 7.13, a thermal image of carbon fibre bundles in a three-dimensional carbon–carbon composite sample, illustrates the high resolution that has been achieved. This is a modulated optical reflectance phase image obtained using a pump laser with~1 μm beam radius, modulated at a frequency of 2 MHz. The contrast between the fibre bundles and the matrix originates in the order of magnitude difference in thermal diffusivity of the two components.

7.3 Time-domain, transient thermal NDE

The chief advantage of time-domain photothermal techniques is that a full photothermal response spectrum of a test sample is obtained, eliminating the need to establish the optimum operating frequency necessary with single frequency cw techniques. The recorded spectrum, a thermal transient, contains complete information about the underlying thermal structure whereas the cw techniques only provide information corresponding to a particular thermal wave frequency. Time-domain measurements can be used for: assessing coating thicknesses, locating adhesion defects and delaminations and detecting local structure variations. They may also be used for the determination of the thermal parameters (as discussed in Chapter 9). In the following sections we will present NDE applications in a format similar to that used above for the frequency-domain photothermal NDE applications. The theoretical background for transient thermal NDE will be found in Chapter 5. The review will be limited to photothermal techniques in which a stationary sample is excited by an optical pulse after which the resulting thermal transient is recorded.

7.3.1 Transient photothermal infrared NDE techniques

Of all the photothermal detection methods, discussed in Chapter 4, that might be used for thermal transient detection, only infrared detection has progressed from the laboratory to provide the basis of new NDE techniques that are acceptable to industry. Two classes of photothermal instruments have been developed: single spot heating with single spot detection and line or area heating with line or area detection. In single spot heating and detection instruments, the system records synchronously the variation of the emitted infrared signal amplitude with time. For the line or area heating and detection instruments there is a sampling of emitted thermal radiation from specific localized regions on the sample surface by an array or scanning detector.

Pulses of thermal energy can be applied by: flash lamps; air-jets or a pulsed laser. The point sensing techniques have a higher temperature and time resolution than the techniques that employ array and scanning detectors. The line and area scanning systems, however, have the great advantage of providing rapid inspections of large sample areas since both temporal and spatial information is obtained at the same time. The immediate availability of images can provide an almost instantaneous indication of the presence or absence of defects in a component. An area scanning system most often takes the form of an infrared camera to record thermal radiation from a sample surface at video frame rates, 25 Hz, following thermal excitation. The integration of computers with infrared cameras has enabled thermography to become a fully quantitative NDE technique. Variations in the sub-surface structure impede the normal flow of heat through the sample and appear as local hot spots in an infrared image. In coated systems, changes in thickness may be revealed as steps or gradients in signal amplitude.

7.3.2 Coating thickness measurements

The thermal transients obtained from coated samples were dealt with theoretically in Section 5.3. It was shown that the transient response to a pulse excitation deviated from a characteristic $t^{-0.5}$ decay at a time determined by coating thickness $t = L^2/\alpha$. This provides a means of coating thickness assessment if thermal diffusivity α is known or if calibration is possible. Figure 7.14 shows the signal transient response to step heating of a set of different thickness zirconia thermal barrier coatings on a stainless steel substrate. The transients clearly differ in a systematic fashion with coating thickness. Coating thickness may be obtained by data fitting to an appropriate model or by a calibration procedure. Some

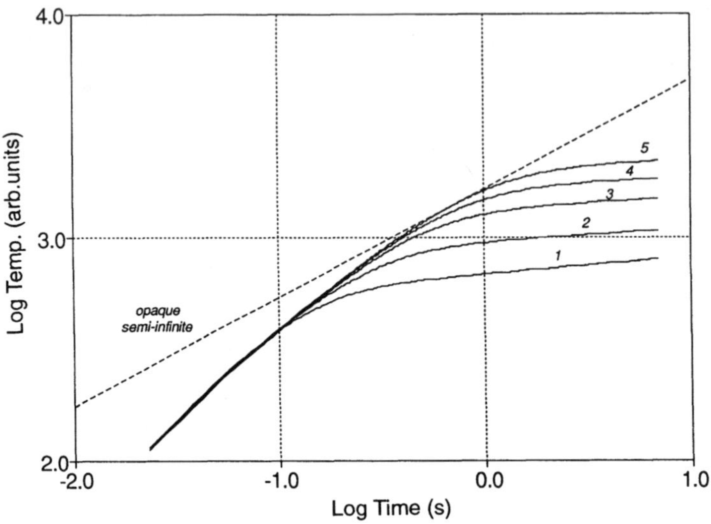

Figure 7.14 *Experimental thermal transients obtained from step heating yttria stabilized zirconia coatings on stainless steel plates. Coating thicknesses were (1) 250, (2) 355, (3) 500, (4) 735 and (5) 850 μm.*

appropriate expressions for data fitting which were given in Chapter 5 are shown in Table 7.2.

The time dependence of data is usually analyzed as this is unaffected by surface optical absorption and emissivity variations (the equivalent of the choice of phase in thermal wave studies).

The effects of coating translucency are illustrated in Figs 7.15(a) and 7.15(b) where the slopes (normalized first time derivative, equation (5.14)) of data for a thermal barrier zirconia coating on steel are shown. These two figures compare the effects of coating translucency and opaqueness on transients obtained from coatings of different thicknesses, in response to step heating. At long times, saturation sets in because of the finite heat spot size and because surface heat loss by convection becomes significant (see Section 5.7). However, at short times the effects of coating thickness variation can be readily resolved. Translucency again has the advantage of extending the range of coating thicknesses that can be assessed, due mainly to the sub-surface generated thermal effects. Coating thickness in a particular range can be obtained from a calibration of the logarithmic thermal transient slopes at a selected time against

Table 7.2 Expressions for data fitting

System	Surface temperature
Step pulse	$T(0,t)=\frac{(1-r)Q_0}{2\sqrt{\pi\rho ckt}}\left[1+2\sum_{n=0}^{\infty}R^n\exp(-n^2L^2/\alpha t)\right]$
Opaque film/ substrate	$t(t)=\frac{2(1-r)Q_0}{k}\sqrt{\frac{\alpha t}{\pi}}\left[1+2\sqrt{\pi}\sum_{n=0}^{\infty}R^n\mathrm{ierf}\,c\left(\frac{L}{\sqrt{\alpha t}}\right)\right]$

R = thermal wave reflection coefficient, L coating thickness, r optical reflection coefficient, Q_0 incident optical intensity

known coating thicknesses. Figure 7.16 shows calibration curves at three times obtained from the data shown in Fig. 7.15(b). The time selected determines the range of thicknesses that can be assessed.

The suitability of transient techniques for coating thickness measurement depends on the magnitude of the effect of a coating on a transient which is determined by the thermal mismatch between the coating and substrate materials. The technique is based on the analysis of transients which is not as simple as the measurement of phase angle in the equivalent cw technique. In addition, the signal-to-noise ratio of the cw technique is intrinsically higher because the thermal energy is confined to a single thermal wave frequency which can be chosen to maximize interference in the coating. For transient techniques, interference occurs with only the fraction of the Fourier components of the excitation pulse that satisfy the interference conditions.

Figure 7.17 shows transients calculated for two coating materials having a large and a moderate thermal mismatch (thermal wave reflection coefficients – 0.6 and –0.3) with a common stainless steel substrate. Whilst both mismatches have proved adequate for the cw thermal wave technique to produce good coating thickness discrimination, the very small influence of the nickel–chrome carbide coatings, Fig. 7.17(b), on transients has proved very difficult to detect experimentally.

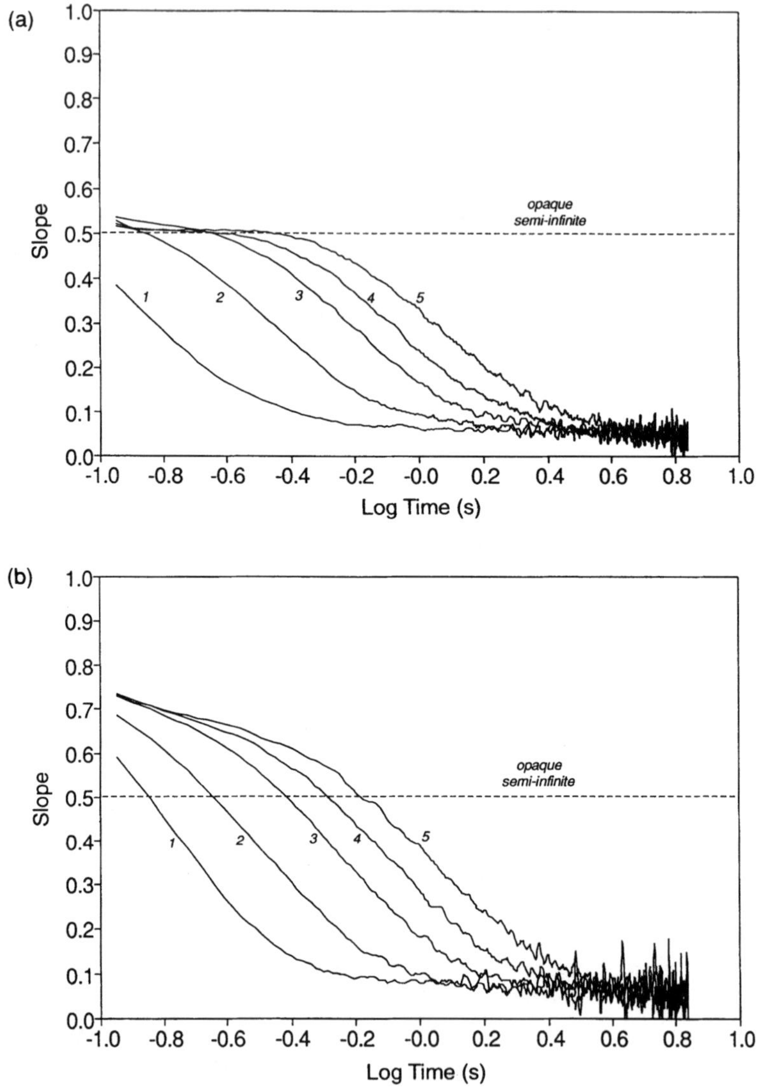

Figure 7.15 *Logarithmic slopes of experimental thermal transients obtained from step heating yttria stabilized zirconia coatings on stainless steel plates: (a) surface darkened and (b) undarkened. Coating thicknesses were (1) 250, (2) 355, (3) 500, (4) 735 and (5) 850 μm.*

7.3.3 Transient response of sub-surface defect

The transient response of a solid is affected by the presence of air gap (crack-like) or thermal contact resistance defects. Figure 7.18 shows calculations of the transients caused by a range of different thickness air gaps between a zirconia coating and a steel substrate, following a heating pulse. In the absence of a defect the transient follows a curve like that shown in Fig. 7.17. The presence of a sub-surface defect which impedes normal heat flow leads to a local increase in the sample surface temperature, or 'thermal contrast', causing a reduced decay rate. The onset of the deviation from the defect-free locus will occur at a characteristic thermal propagation time, L^2/α. The magnitude of the thermal contrast depends on the thickness of the air-gap or on the magnitude of the thermal contact resistance, in much the same way as was discussed earlier for thermal wave phase changes at defects.

A comparison of the variation of contrast with air gap thickness obtained experimentally and theoretically is given in Fig. 7.19. The results were obtained from the surface of a 0.46 mm thick epoxy block adjusted over a thick mild steel block to sandwich a wide range of air-gap thicknesses.

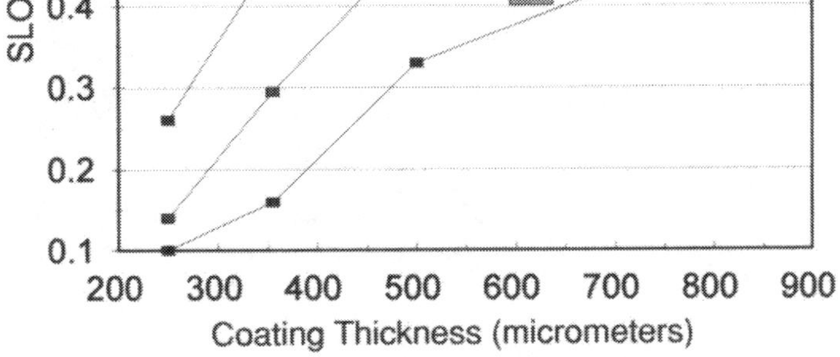

Figure 7.16 *Calibration graphs of logarithmic slopes from Fig. 7.15 (b) versus coating thickness measured at (1) $10^{-0.6}$, (2) $10^{-0.4}$ and (3) $10^{-0.2}$ s.*

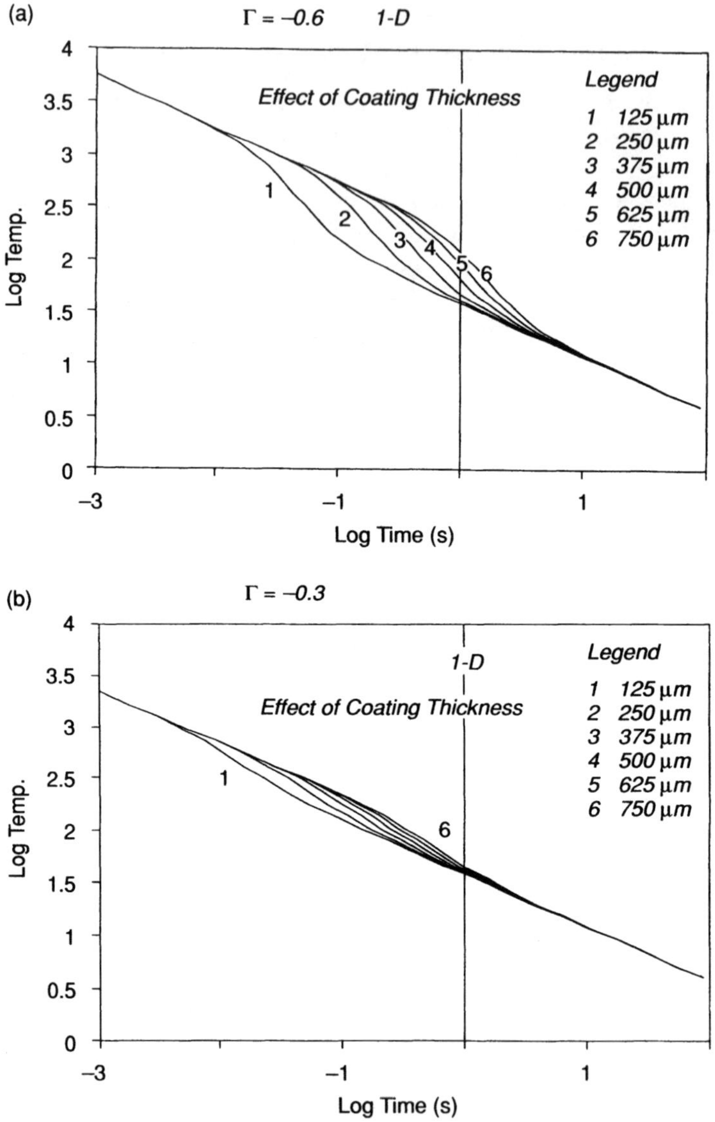

Figure 7.17 *Pulse excitation thermal transients calculated for (a) yttria stabilized zirconia and (b) nickel–chrome carbide coatings on stainless steel substrates.*

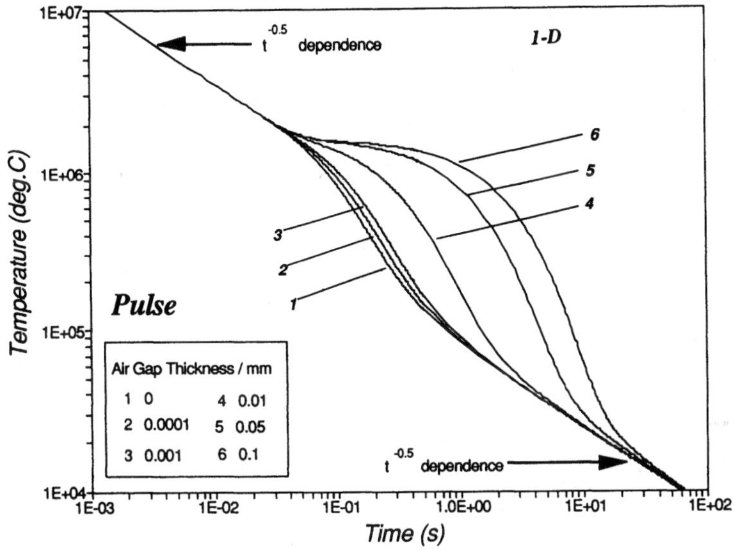

Figure 7.18 *Pulse excitation thermal transients calculated for the air gaps indicated between yttria stabilized zirconia coatings, thickness 0.25 mm, and stainless steel substrates.*

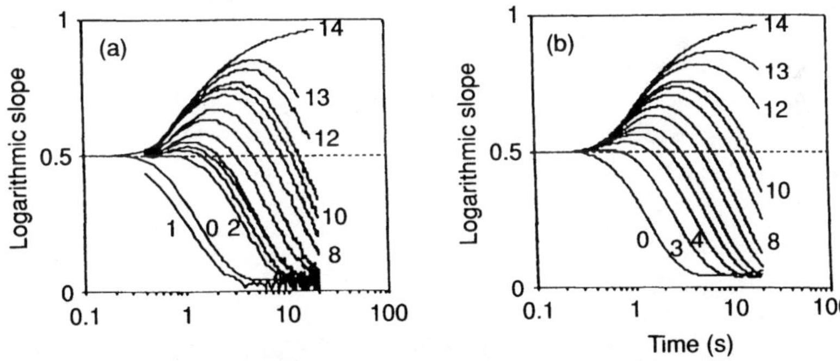

Figure 7.19 *Experimental (a) and theoretical (b) logarithmic slopes of thermal transients for a 0.46 mm thick epoxy plate on a mild steel block sandwiching air-gap thickness: (13) 2 mm; (12) 1 mm; (11) 0.5 mm; (10) 0.4 mm; (9) 0.3 mm; (8) 0.2 mm; (7) 0.15 mm; (6) 0.1 mm; (5) 75 μm; (4) 50 μm; (3) 25 μm; (2) 0 μm. (0) and (14) are theoretical responses for perfect contact between plate and block and for an air backed plate, respectively (S.K. Lau, unpublished).*

As for the thermal wave case, defect effects also depend on the thermal properties of the host material. The total thermal contrast also depends on the lateral extent of the defect concerned and will, in general, be smaller than shown in Figs 7.18 and 7.19 which were calculated assuming infinite lateral extent. Consequently, whilst it is feasible to obtain the characteristics of a defect from the magnitude of the contrast it produces, care must be taken to account for its finite lateral extent.

7.3.4 Transient thermography

A transient thermography NDE system, Fig. 7.20, comprises a broad area heat source, usually a powerful flash lamp, and an infrared camera. The source and camera are synchronized to enable the camera to record the transient thermal decay of a flash heated surface. Artificial defects in the form of flat bottom channels, flat bottom cylindrical holes, air-gap disk defects, cylindrical and spherical holes, and strip and disk contact resistances have been readily imaged by transient thermography. Transient thermography has also been applied: to detect real adhesive bonding defects in aluminum lap joints typically used in aircraft skin structures; to detect corrosion in pipes; to detect faults in composite material panels; to detect defects below thick epoxy coatings on steel tanks and for the mapping of thermal diffusivity variations. A transient thermography image of artificial defects in a carbon composite panel is shown in Fig. 7.21. Although an image of similar resolution of the same component could have been produced by ultrasonic c-scan, it would

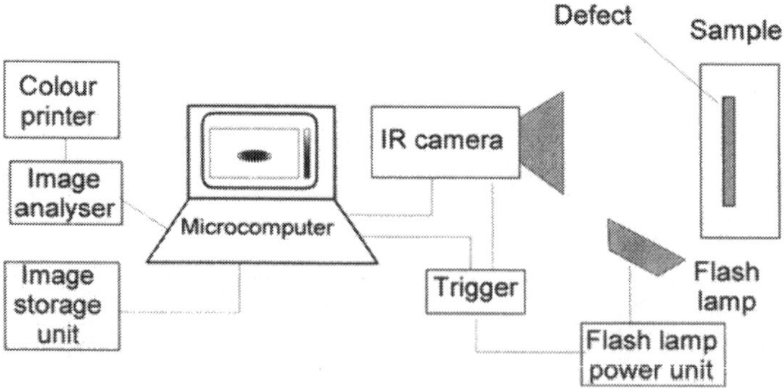

Figure 7.20 *Block diagram of a transient thermography system.*

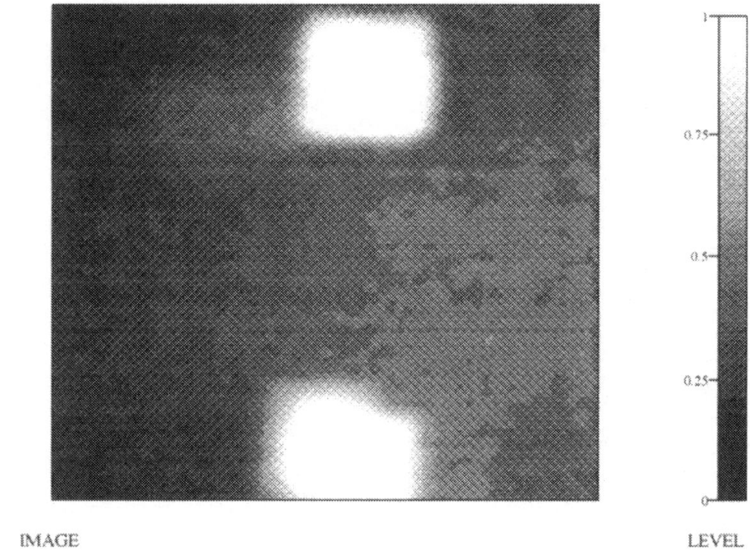

Figure 7.21 *Transient thermography image of two artificial defects in a carbon fiber composite panel. The defects were 25 mm × 25 mm squares of Teflon inserted between laminates in place of carbon fiber (provided by P. Delpech).*

have taken a great deal longer to obtain and the panel would have needed to be immersed in a liquid, usually water, filled scanning tank.

In comparison with conventional ultrasonic or radiographic testing, transient thermography has the advantages of:

1. fast scanning for imaging;
2. ease of deployment;
3. a safe mode of operation;
4. being single sided, access only required to one side of a component;
5. being closely integrated to computer based systems.

Several analytical methods have been proposed to obtain quantitative defect information from transient thermographic images. For a planar sub-surface defect such as an air-gap or contact resistance, a data fitting procedure can be employed to characterize the defect (depth, thickness, thermal contact resistance). However, this type

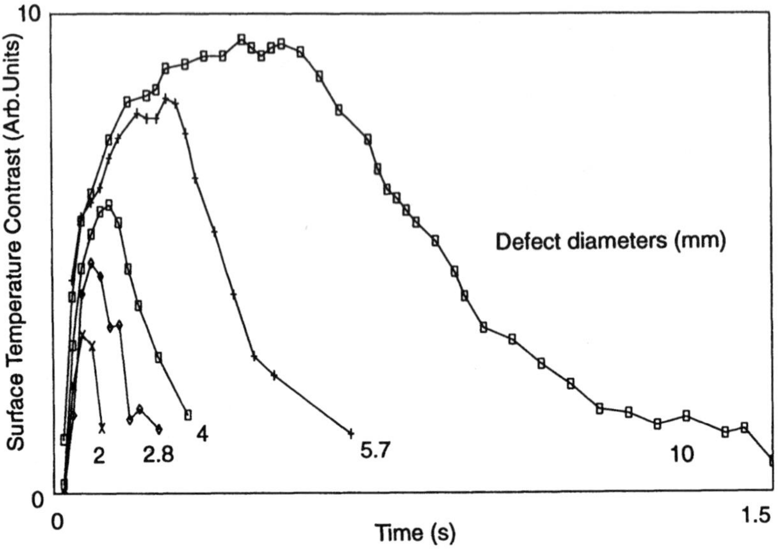

Figure 7.22 *Transient thermography thermal contrast–time dependences of a range of different diameter circular defects at a depth of 1 mm in mild steel.*

of data fitting is only possible if the thermal properties of the layers involved are known.

An alternative procedure is to calibrate the time for maximum contrast of a defect image, obtained from a contrast time graph. This parameter varies with the square of the defect depth, provided the defect is reasonably large and simple one-dimensional transient heat flow relationships hold. Such a calibration can be used to determine the depths of real defects or to construct images of defect depth rather than temperature amplitude. This analysis is the basis of thermal tomography in which a sequence of thermographic images is analyzed to obtain a single timeogram image which reflects the depth of a sub-surface structure. Vavilov *et al.* (1993) have demonstrated the power of this technique in the analysis of images of defects in carbon epoxy composites.

Timeogram images must be interpreted with care because the time for maximum contrast varies with defect width. This invalidates the above relationship with defect depth for small defects and distorts timeogram images. Experimental data for a set of different diameter circular defects, all at the same depth, in a mild steel

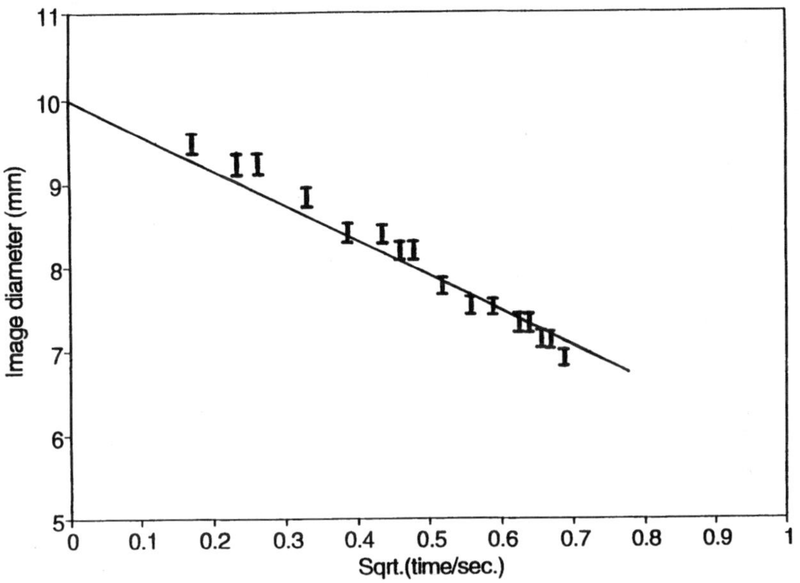

Figure 7.23 *The shrinkage of the widths (FWHM) of transient thermographic images of a 10 mm circular defect at a depth of 1 mm beneath the surface in a block of mild steel.*

sample shown in Fig. 7.22 illustrate this effect very well. It is clear that, for small defects, the time for maximum contrast is strongly dependent on defect size and that these times would not be a good measure of defect depth. The initial slopes of the curves, however, are a common feature that might be used to assess defect depth.

The lateral extent of a defect can be obtained by making use of a sequence of its thermal images, recorded at the surface. The widths of a defect's thermal images have been found to shrink systematically with elapsed time, as shown in the example in Fig. 7.23. An accurate indication of defect size can be obtained by extrapolating these image widths back to the pulse excitation time. To use this technique it is necessary to obtain thermal contrast profiles of the defect images and to use the full widths at half maximum amplitude as measures of image width.

A major attraction of transient thermography is the close integration of infrared cameras with increasingly powerful computer systems. There is great scope for image processing to enhance the images of defects, to correct for inherent limitations in spatial or thermal imaging resolution and to perform analytical algorithms that provide quantitative defect information.

References

Grice, K.R., Inglehart, L.J., Favro, L.D., Kuo, P.K. and Thomas, R.L., 'Thermal wave imaging of closed cracks in opaque solids', *J. Appl. Phys.*, **54** (1983) 6245–6255.

Thomas R.L., Pouch J.J., Wong Y.H., Favro LD and Kuo P.K., 'Subsurface flaw detection in metals by photoacoustic microscopy', *J. Appl. Phys.* **51** (1980) 1152–56.

Vavilov, V., Maldague, X., Dufort, B., Robitaille, F. and Picard, J., 'Thermal nondestructive testing of carbon epoxy composites: detailed analysis and data processing', *NDT & E Int.*, **26** (1993) 85–95.

Further reading

Aamodt, L. C. and Murphy, J C., 'Effect of 3-D heat flow near edges in photothermal measurements', *Appl. Optics*, **21** (1982) 111–115.

Almond, D. P. and Lau, S. K., 'Edge effects and a method of defect sizing for transient thermography', *Appl. Phys. Lett.*, **62** (1993) 3369–3371.

Busse, G., 'Optoacoustic and photothermal material inspection techniques', *Appl. Optics*, **21** (1982) 107–110.

Busse, G. and Rosencwaig, A., 'Subsurface imaging with photoacoustics', *Appl. Phys. Lett.*, **36** (1980) 815–16.

Favro, L. D., 'Progress in the development of pulse-echo thermal wave imaging for NDE', in *Rev. Prog. Quantitative NDE*, **13A** (eds D. O. Thompson and D. E. Chimenti) Plenum Publishing (1994) 395–400.

Grice, K. R., Favro, L. D., Kuo, P. K. and Thomas, R. L., 'Evaluation of photoacoustic microscopy fatigue crack detection', in *Rev. Prog. Quantitative NDE*, **2B** (eds D. O. Thompson and D. E. Chimenti) Plenum Publishing (1983) 1019–27.

Hartikainen, J., Lehtiniemi, R., Rantala, J., Varis, J. and Luukkala, M., Fast infrared line-scanning method and its applications', *Rev. Prog. Quantitative NDE*, **13A** (eds D. O. Thompson and D. E. Chimenti) Plenum Publishing (1994) 401–408.

Imhof, R. E., Zhang, B., and Birch, D. J. S., 'Photothermal radiometry for NDE', in *Progress in photothermal and photoacoustic science and technology* 2 (ed. A. Mandelis), Elsevier, New York (1993) 185–236.

Inglehart, L. J., Grice, K. R., Favro, L. D., Kuo, P. K. and Thomas, R. L., 'Spatial resolution of thermal wave microscopes', *Appl. Phys. Lett.*, **43** (1983) 446–8.

Lau, S. K., Almond, D. P. and Patel, P. M., 'Transient thermal wave techniques for the evaluation of surface coatings', *J. Phys. D: Appl. Phys.*, **24** (1991) 428–36.

Luukkala, M., Lehto, A., Jaarinen, J. and Jokinen, M., 'Photothermal imaging and thermal surface waves as a NDT tool for coatings', *Proc. IEEE Ultrasonics Symposium* (1982) 591–94.

Murphy, J. C., Spicer, J. W. M., Osiandra, R., Kerns, W. D. and Aamodt, L. C., 'Quantitative nondestructive evaluation of coatings by thermal wave imaging', in *Rev. Prog. Quantitative NDE*, **13A** (eds D. O. Thompson and D. E. Chimenti) Plenum Publishing (1994) 417–426.

Opsal, J. and Rosencwaig, A., 'Thermal-wave depth profiling: Theory', *J. Appl. Phys.*, **53** (1982) 4240–4246.

Opsal, J., Rosencwaig, A. and Willenborg, D. L., 'Thermal wave detection and thin-film thickness measurement with laser beam deflection', *Appl. Optics*, **22** (1983) 3169–76.

Pade, O. and Mandelis, A., 'Computational thermal-wave slice tomography with back propagation and transmission reconstructions', in *Rev. Prog. Quantitative NDE* **13A** (eds D. O. Thompson and D. E. Chimenti) Plenum Publishing (1994) 409–416.

Patel, P. M., Almond, D. P. and Reiter, H., 'Thermal wave detection and characterisation of subsurface defects', *Appl. Phys.*, **B43** (1987) 9–15.

Patel, P. M., Almond, D. P. and Morris, J. D., 'Potential applications of thermal wave interferometry for non-destructive evaluation and characterisation of surface coatings', *European J. NDT*, **1** (1991) 64–70.

Rosencwaig, A., Opsal, J. and Willenborg, D. L., 'Thin-film thickness measurement with thermal waves', *Appl. Phys. Lett.*, **43** (1983) 166–8.

Tam, A. C. and Sontag, H., 'Measurement of air gap thickness underneath an opaque film by pulsed photothermal radiometry', *Appl. Phys. Lett.*, **49** (1986) 1761–63.

8

Thermal wave characterization of semiconductors

The characterization of semiconductor materials is an important and successful application of photothermal science. In particular, the modulated optical reflectance (MOR) technique has been demonstrated to provide an attractive non-contact method for the non-destructive evaluation of semiconductor wafers. This technique has been shown to be effective for monitoring a variety of wafer processing stages (polishing, scrubbing, etching and doping) and for rapidly identifying variations in processing performance. This ability to monitor each stage of wafer processing can lead to the identification of the parameters which critically affect wafer quality. For these reasons MOR systems have been developed commercially, specifically for the semiconductor industry.

In this chapter we examine the capabilities of some of the photothermal techniques (PA, MOR, and OBD) for semiconducting material characterization. The characterizations considered are: the general monitoring of wafer quality; the determination of electronic parameters (carrier diffusivity and lifetime), and the measurement of doping and of the dimensions of implanted regions. An outline of the general influence of the semiconductor optical properties on their photothermal response can be found in Chapter 3.

8.1 Photothermal response of semiconducting materials

In Chapters 2 and 3 it was assumed that heat generation by light absorption was localized in the region where the absorption occurred

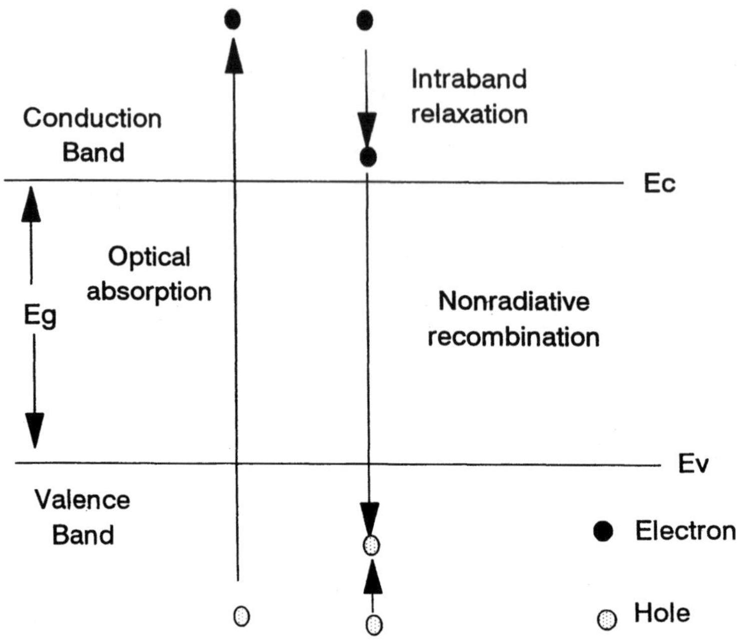

Figure 8.1 *Schematic diagram of optical absorption processes in a semiconductor.*

and that it was instantaneous. The latter assumption arose from the recognition that optical absorption, in most materials, leads to atomic or molecular excitations which are short-lived, the excitation energy being released to the lattice as heat on a time scale that is short when compared with the modulation periods employed in photothermal experiments. These assumptions, however, are not valid for semiconducting materials. When a semiconductor is irradiated by light with photon energy $E = h\nu$ larger than the bandgap E_g, appropriate for optical absorption, electrons are excited from the valence band to energy states $E - E_g$ higher than the lowest states of the conduction band, leaving holes in the valence band, as illustrated in Fig. 8.1. This excess energy $E - E_g$ is transferred to the lattice, on a picosecond time scale, as the excited electrons drop to the bottom of the conduction band. This heating of the lattice forms the primary thermal wave source. After a much longer characteristic time, the carrier lifetime τ, the remaining energy E_g is dissipated to the lattice, mostly in the form of heat, when the excited carriers in the conduction band recombine with the holes in the valence band. During the time τ the electrons diffuse into the material thereby moving the energy from the point of excitation, causing a secondary thermal wave source that has a spatially extended distribution and which develops over a time scale τ after the primary source. In addition, some of the excited electrons will diffuse to the surface where they recombine very rapidly because of the high density of carrier traps and surface states present at the surface. This process produces a third heat source at the surface.

8.2 Carrier and temperature distributions

The photothermally induced carrier and temperature distributions in semiconductors depend on both carrier and thermal diffusion processes. These distributions may be obtained from the appropriate carrier and thermal diffusion equations (neglecting acoustic waves, thermomechanical and electromechanical coupling contributions). To introduce principles and basic phenomena a simple one-dimensional treatment will be given here, whilst noting that in practice the use of highly focused probes will necessitate the development of a full three-dimensional treatment. The one-dimensional Fourier equation (2.16) that models thermal diffusion in the bulk takes the modified form:

$$\alpha \frac{\partial^2 T}{\partial z^2} = \frac{\partial T}{\partial t} - \frac{E_g \alpha}{k} \frac{N(z,t)}{\tau} \tag{8.1}$$

in which the second term on the right accounts for the secondary heating caused by the recombinations of the diffusing carriers, after a characteristic time τ; $N(z, t)$ is their distribution. Assuming the semiconductor to absorb light at the surface, the boundary conditions take the form:

$$-k \left. \frac{dT}{dz} \right|_{z=0} = \frac{(h\nu - E_g) I_0}{2h\nu} e^{j\omega t} + sN(0,t)E_g \tag{8.2}$$

in which the first term on the right is the primary thermal wave source, I_0 is the absorbed optical intensity, and the second term describes the surface recombination in terms of the surface recombination velocity s. The excited carrier density distribution is given by a modified diffusion equation:

$$D\frac{\partial^2 N}{\partial z^2}=\frac{\partial N}{\partial t}+\frac{N}{\tau} \tag{8.3}$$

in which the second term on the right accounts for the decay in carrier concentration caused by recombinations in the bulk, D is the carrier diffusivity. At the surface, carriers are both created by optical absorption and eliminated by surface recombinations. The surface boundary conditions are:

$$D\frac{dN}{dz}\bigg|_{z=0}=-\frac{I_0}{2h\nu}\mathrm{e}^{\mathrm{j}\omega t}+sN(0,t) \tag{8.4}$$

The solution for the excited carrier density distribution is:

$$N(z,t)=\mathrm{Re}\left[\frac{I_0}{2h\nu}\left(\frac{\mathrm{e}^{-pz}}{pD+s}\right)\mathrm{e}^{\mathrm{j}\omega t}\right] \tag{8.5}$$

where $p = \surd[(1+\mathrm{j}\omega\tau)/(D\tau)]$. In the limit $\omega\tau \ll 1$, where the photothermal modulation frequency is very low compared with the carrier recombination rate, the carrier concentration adopts a simple diffusive decay with depth z. It loses its dependence on modulation frequency and has a conventional diffusion length of $\surd(D\tau)$. When $\omega\tau \gg 1$, the photothermal modulation frequency is much greater than the recombination rate and the carrier distribution becomes wave-like. A plasma wave is formed in which the carrier density has wave-like properties similar to those of temperature in a thermal wave. The carrier density varies periodically with depth but is exponentially damped in amplitude. In this limit, a plasma diffusion length, $\mu_p = \surd(2D/\omega)$, may be defined that is analogous to the thermal diffusion length. One essential difference between a plasma wave and a thermal wave, however, is the effect of carrier recombinations, characterized by the recombination time.

The solution for the temperature distribution can be written as:

$$T(z,t)=\frac{I_0}{2k\sigma}\times$$

$$\left[(1-\varepsilon)\mathrm{e}^{-\sigma.z}+\frac{\varepsilon(p\mathrm{e}^{-\sigma.z}-\sigma\mathrm{e}^{-p.z})}{D\tau(p+s/D)(p^2-\sigma^2)}+\frac{s\varepsilon\mathrm{e}^{-\sigma.z}}{(pD+s)}\right]\mathrm{e}^{\mathrm{j}\omega t} \tag{8.6}$$

where $\varepsilon = E_g/h\nu$ and σ is the thermal wave number. The above expression is written in a form which shows explicitly the respective contributions of the three heat source terms. The first term in the large square bracket is the contribution of the instantaneous intraband non-radiative thermallizations that form the primary source; the second term is the interband non-radiative bulk recombinations and the third results from the non-radiative surface recombinations. In the absence of a bandgap, energy $E_g = 0$, the temperature distribution becomes that of a simple thermal wave.

The dependence of temperature and carrier concentration on the parameters D, s and τ are illustrated in Figs 8.2 and 8.3. The computations, by Liu *et al.* (1993), are for bulk silicon illuminated by a focused 20 mW laser beam, Gaussian radius 4 μm, modulated at a frequency of 100 kHz. The carrier concentration and temperature at the beam center, Fig. 8.2, are found to depend on both carrier lifetime and surface recombination velocity. Carrier concentration at the surface falls with surface recombination velocity and surface temperature increases with the increase in carrier recombinations at the surface. The dependence on carrier lifetime splits into two regimes: (1) $\omega\tau \ll 1$ and (2) $\omega\tau \gg 1$. When $\omega\tau \ll 1$, τ is very small ($\ll$1 μs) most of the recombinations of the carriers created at the surface occur close to the surface, reducing surface carrier concentration and raising surface temperature. When $\omega\tau \gg 1$ ($\tau \gg$1 μs) the recombination rate at the surface is lower, surface carrier concentration is higher and surface temperature is smaller for the lower values of surface recombination velocity. Typical values for a silicon sample are: $\tau > 1$ μs and $s < 100$ m s^{-1}. The influence of carrier diffusion coefficient D on carrier concentration and temperature is shown in Fig. 8.3. Whilst temperature is insensitive to D, carrier concentration falls as D^{-1}.

Figure 8.2 *Surface carrier concentration (a) and surface temperature (b) as a function of surface recombination velocity and carrier lifetime for bulk silicon illuminated by a 20 mW laser beam modulated at 100 kHz (Liu et al., 1993; reproduced with permission).*

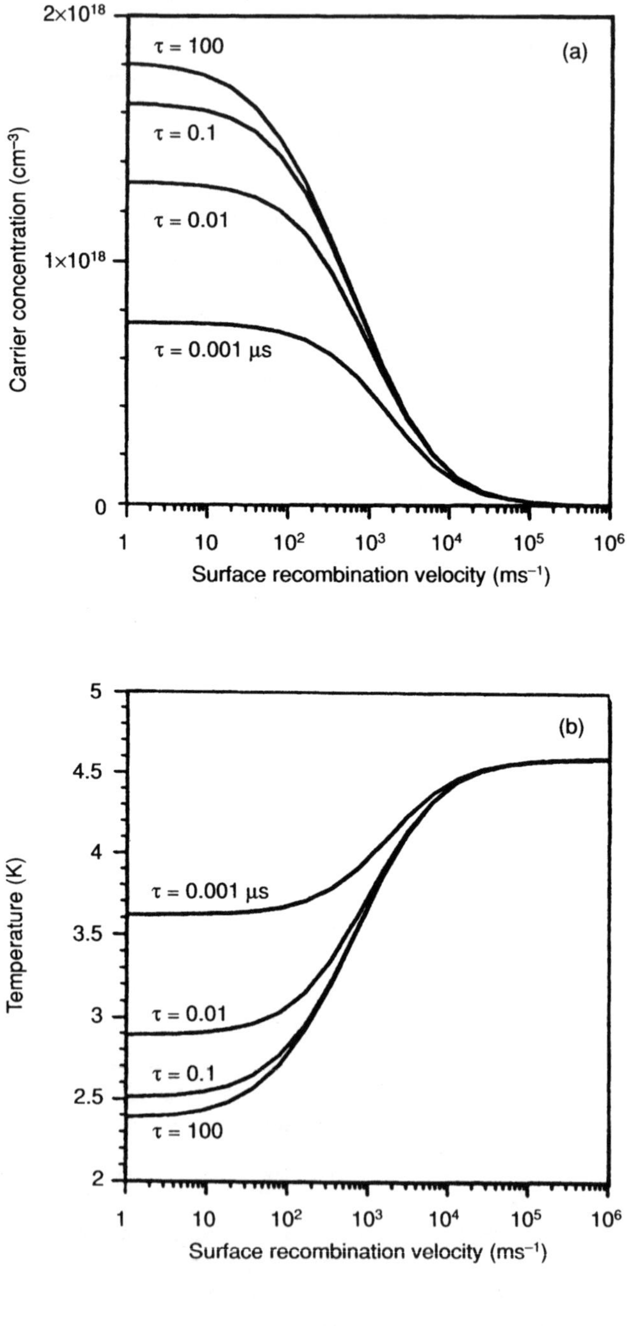
2×10^18
1×10^18
0
Carrier concentration (cm^-3)
τ = 100
τ = 0.1
τ = 0.01
τ = 0.001 μs
(a)
1
10
10^2
10^3
10^4
10^5
10^6
Surface recombination velocity (ms^-1)
5
4.5
4
3.5
3
2.5
2
Temperature (K)
(b)
τ = 0.001 μs
τ = 0.01
τ = 0.1
τ = 100
1
10
10^2
10^3
10^4
10^5
10^6
Surface recombination velocity (ms^-1)

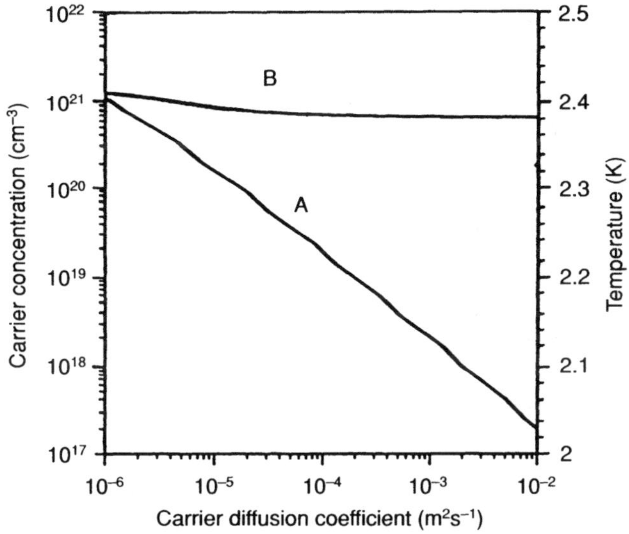

Figure 8.3 *Surface carrier concentration (A) and temperature (B) as a function of carrier diffusion coefficient for bulk silicon illuminated by a focused 20 mW laser beam modulated at 100 kHz (Liu et al., 1993; reproduced with permission).*

The above carrier and temperature distributions were derived assuming that the semiconductor sample was opaque and thermally thick. If the sample has finite thickness and optical absorption coefficient, the distributions are characterized by three additional parameters: the sample thickness; the optical absorption coefficient and the rear face surface recombination velocity.

8.3 Photoacoustic signal

In Chapter 4 it was noted that a photoacoustic signal is directly proportional to sample surface temperature. Hence the photoacoustic effect can, in principle, be used to make frequency response studies from which the electronic properties (D, τ and s) of a semiconductor can be obtained, provided the thermal properties of the sample are known accurately. Figure 8.4 shows calculations of the frequency

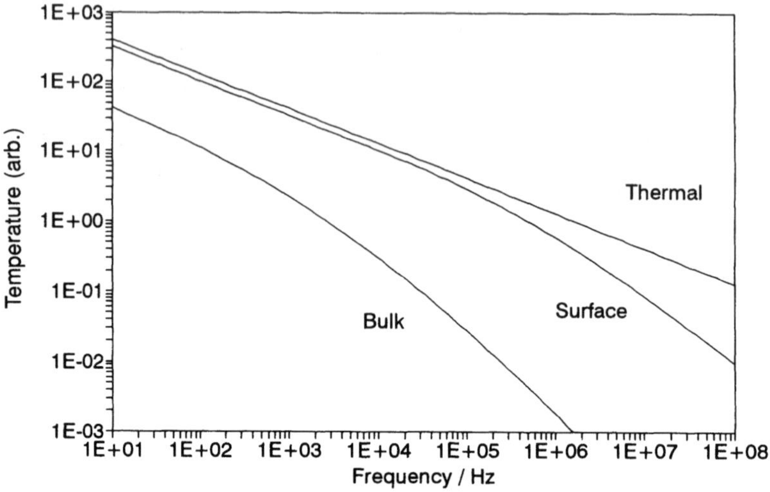

Figure 8.4 *Frequency dependences of the three components in equation 8.6 to the surface temperature of bulk crystalline silicon, calculated taking* $s = 100\ m\ s^{-1}$, $D = 20\ cm^2\ s^{-1}$ *and* $\tau = 10^{-5}\ s$.

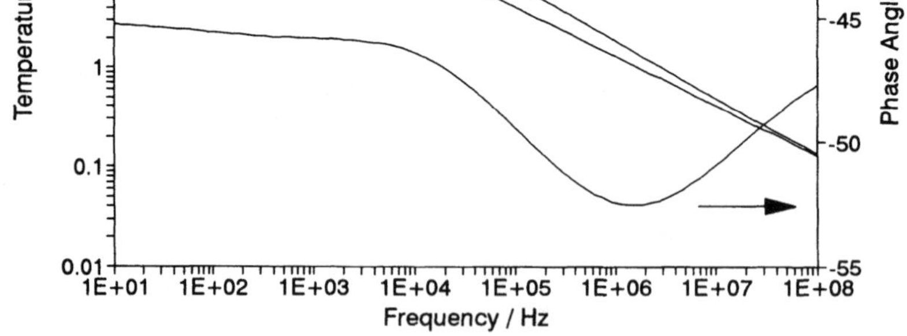

Figure 8.5 *Frequency dependence of photoacoustic amplitude and phase for bulk crystalline silicon, calculated taking* $s = 100\ m\ s^{-1}$, $D = 20\ cm^2\ s^{-1}$ *and* $\tau = 10^{-5}\ s$.

dependences of the three contributions in equation (8.6) to the surface temperature of bulk crystalline silicon. All three diminish with frequency and the enhanced frequency dependences, at higher frequencies, of the surface ($f^{-1.0}$) and bulk terms ($f^{-1.5}$) arise from the significant increase in the factor p with frequency when the

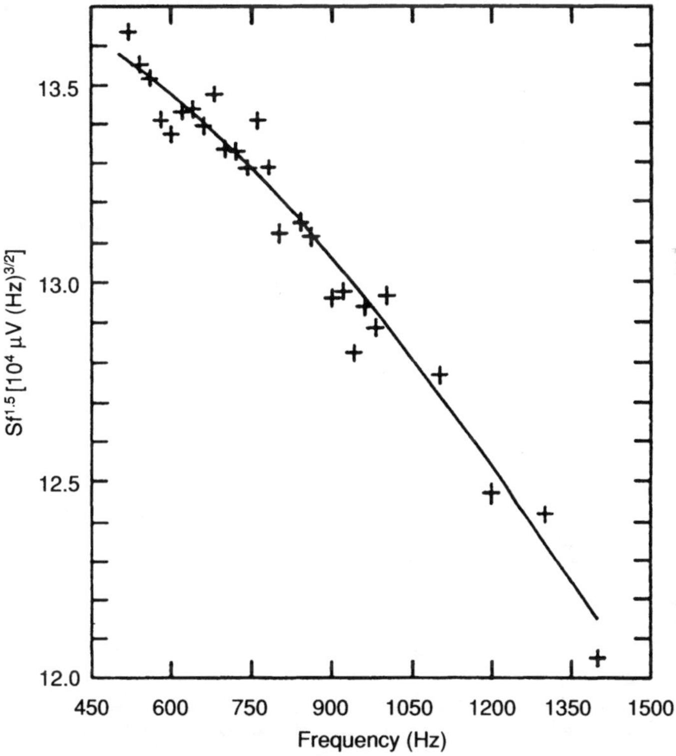

Figure 8.6 *The photoacoustic signal obtained at the rear face of a silicon sample* $\times f^{1.5}$ *as a function of modulation frequency. The solid line is a theoretical fit to the data. (Reproduced with permission from Pinto Neto et al., 1990. Copyright 1990 The American Physical Society.)*

condition $\omega\tau \gg 1$ is satisfied. This factor also develops a significant imaginary component in the same frequency range which causes the changes in phase shown in Fig. 8.5. However, this distinctive free carrier effect feature occurs at frequencies above 50 kHz, which exceeds the frequency range for photoacoustic detection. At lower frequencies the photoacoustic response is indistinguishable from conventional solids because of the dominance of the thermal term, the first term in equation (8.6).

One solution is to minimize the thermal term by employing a light source of wavelength corresponding to the electronic band gap of the material, i.e. choosing $E_g = h\nu$ so that $\varepsilon = 1$. This effectively eliminates the contribution from the thermal term, leaving the surface temperature determined by the bulk and surface recombination processes. In practice this necessitates the use of a tuneable light source, such as a dye laser.

A much simpler solution is to make use of the exponential decay of the thermal and surface terms with sample depth z. Rear face detection can be employed to monitor the remaining bulk term passing through a thermally thick sample, provided it can be arranged that the parameter $\rho \gg \sigma$ through careful choice of frequency. This ensures that only the term proportional on e^{-pz} in equation (8.6) is detected at the rear face. It has been shown both theoretically and experimentally that this approach can be used to determine the electronic properties of a sample. A typical result is shown in Fig. 8.6 for silicon with roughened front and rear faces. The frequency dependence of the photoacoustic signal is close to $f^{-1.5}$ corresponding to the bulk component. Note the low modulation frequency range employed (up to 1400 Hz) which is well within the capabilities of gas microphone detection.

8.4 OBD (mirage) signal

The infrared transparency/translucency of semiconducting materials provides the opportunity to probe photothermal plasma waves directly using a focused infrared laser beam that passes through the affected region of a sample, as illustrated in Fig. 8.7. The normal beam deflection component assuming plane periodic heating is given by:

$$D_n = \left(\frac{L}{n}\right)\left(\frac{\partial n}{\partial T}\frac{\partial T}{\partial z} + \frac{\partial n}{\partial N}\frac{\partial N}{\partial z}\right) \tag{8.7}$$

in which n is the refractive index and L is the probe path through the sample. The deflection component in this case has a thermal component and a free carrier component, both of which depend on

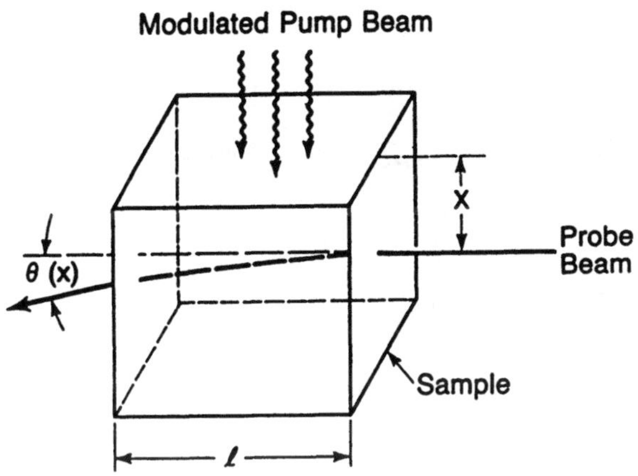

Figure 8.7 *Schematic diagram of OBD (mirage) principle used in the study of 'transparent' semiconductors (Fournier et al., 1986; reproduced with permission).*

the thermal and electronic characteristics of the semiconducting material. The two terms $\partial T/\partial z$ and $\partial N/\partial z$ can be obtained from the expressions (8.5) and (8.6) above.

Figure 8.8 shows a comparison of the experimental and theoretical frequency dependences of the probe beam deflection amplitude and phase from a p-type silicon crystal with the 1.15 μm wavelength probe beam passing through the sample at a range of depths x below the heated surface. The experimental measurements are modelled well by the theory and the parameters D, τ and s obtained by curve fitting agree with other published values. This technique has the advantages of being contactless and of measuring important electronic transport parameters below the surface, within the bulk of the semiconductor, in a spatially resolved manner.

Figure 8.8 *A comparison of the experimental and theoretical frequency dependences of (a) amplitude and (b) the phase of the OBD signal obtained from a p-type silicon crystal with the probe beam passing through the crystal at depths x, indicated in the figure, below the surface. The values* $D = 30\ cm^2\ s^{-1}$ *and* $\tau = 4.5 \times 10^{-5}$ *s were used to generate the fitted curves (Fournier et al., 1986; reproduced with permission).*

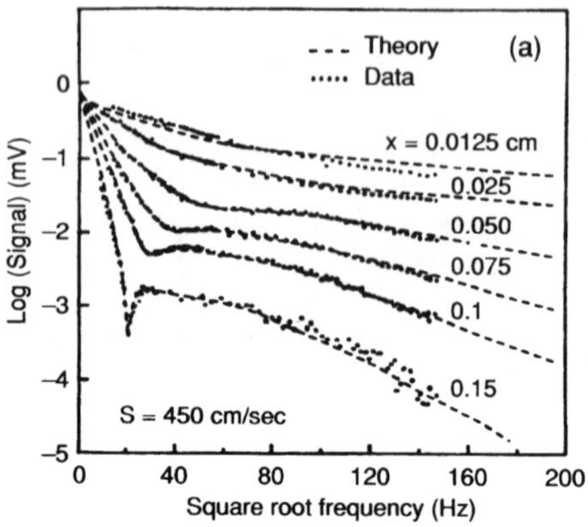

--- Theory
..... Data
(a)
x = 0.0125 cm
0.025
0.050
0.075
0.1
0.15
S = 450 cm/sec
Log (Signal) (mV)
Square root frequency (Hz)

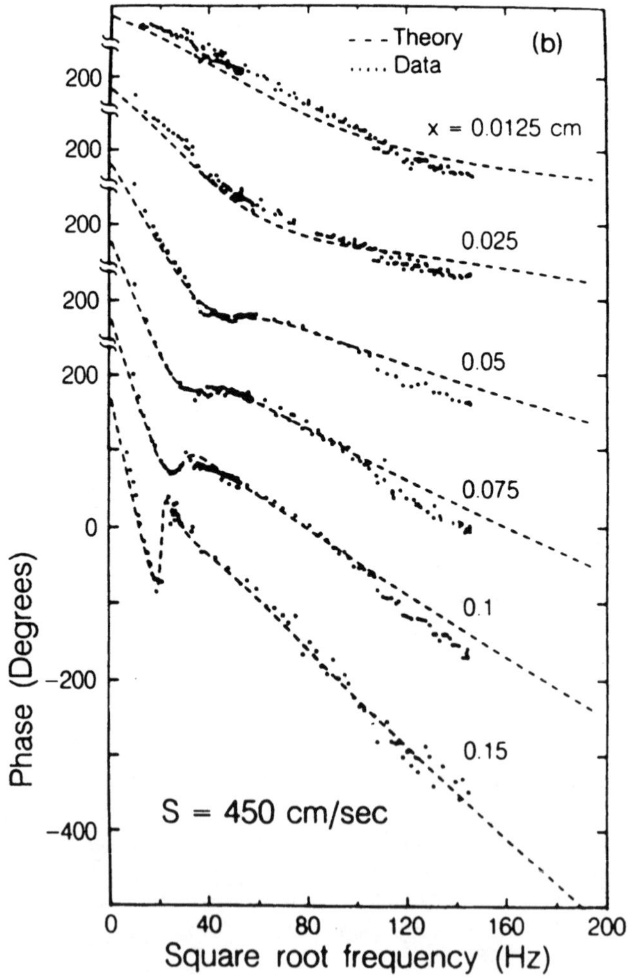

--- Theory
..... Data
(b)
x = 0.0125 cm
0.025
0.05
0.075
0.1
0.15
S = 450 cm/sec
Phase (Degrees)
Square root frequency (Hz)

8.5 Modulated optical reflectance signal

The MOR signal of a semiconductor is determined by the dependence of surface reflectivity R, on two factors: the variation in the optical refractive index with temperature and the change in refractive index caused by carrier generation.

The thermal component is obtained by expressing reflectivity in terms of the complex refractive index, $n' + jn''$, and considering the effect of a small thermal perturbation:

$$\frac{\Delta R_{\text{th}}}{R} = \text{Re}\left[\frac{4T_0}{(n'+\text{j}n''-1)(n'+\text{j}n''+1)}\left(\frac{\partial n'}{\partial T}+\text{j}\frac{\partial n''}{\partial T}\right)\right]$$

in which T_0 is the photothermal change in surface temperature. For crystalline silicon this thermal coefficient of reflectance is estimated to be $+1.5 \times 10^{-4}$ °C^{-1}.

Similarly, the effect on reflectivity of the photothermal perturbation in carrier density N_0 is:

$$\frac{\Delta R_{\text{c}}}{R} = \text{Re}\left[\frac{4N_0}{(n'+\text{j}n''-1)(n'+\text{j}n''+1)}\left(\frac{\partial n'}{\partial N}+\text{j}\frac{\partial n''}{\partial N}\right)\right] \quad (8.9)$$

The coefficients $\partial n'/\partial N$ and $\partial n''/\partial N$ may be obtained from the Drude expressions (3.8) for the complex dielectric constant in the high $\omega\tau$ limit:

$$\frac{\partial n'}{\partial N} = -\frac{\lambda^2 e^2}{8\pi^2 \varepsilon_0 n' m c^2} \quad \text{and} \quad \frac{\partial n''}{\partial N} = -\left(\frac{n''}{n'}\right)\left(\frac{\partial n'}{\partial N}\right) \quad (8.10)$$

in which λ is the wavelength of the reflected light and c is its velocity. For a typical plasma density of 10^{18} cm^{-3} and a probe wavelength of 633 nm this modulated reflectance component is found to be -10^{-4} °C^{-1}, which is of similar magnitude but opposite sign to the above thermal term. Consequently, the partial cancellation of the two terms results in a rather small total modulated reflectance being estimated for silicon. However, the Drude term is linearly proportional to plasma density and this can be increased dramatically by trapping sites at the surface. Defects and lattice damage at a surface have a significant effect on carrier diffusion, effectively pinning the plasma wave to the region just below the surface. As a result, the magnitude of the MOR signal is widely interpreted as providing an indication of surface trap state density or sub-surface damage.

8.6 Studies of semiconductor wafer processing

The ion implantation process that is used to introduce doping atoms into semiconductors also produces lattice disorder and damage. The doping and lattice disorder can affect the local surface temperature, achieved photothermally, in several ways:

1. The optical absorption coefficient is increased, increasing the amount of light absorbed at the sample surface;
2. The local electronic and thermal transport properties are decreased, allowing less heat to be dissipated from the point of excitation;
3. The creation of impurity states within the gap causes an increase in the local non-radiative recombination rate of the photo excited carriers, providing additional heating to the lattice.

Thus the presence of dopant atoms, impurities and defects in an ion implanted layer causes an increase in the local ac temperature

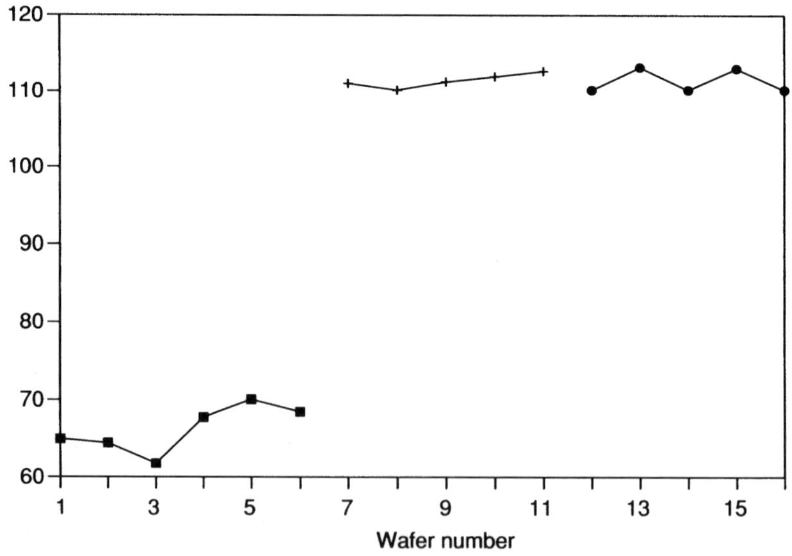

Figure 8.9 *Modulated reflectance signals for scrubbed and non-scrubbed silicon wafers: ■ no scrub,+ brush scrub, ● jet scrub (Rosencwaig, 1987; reproduced with permission).*

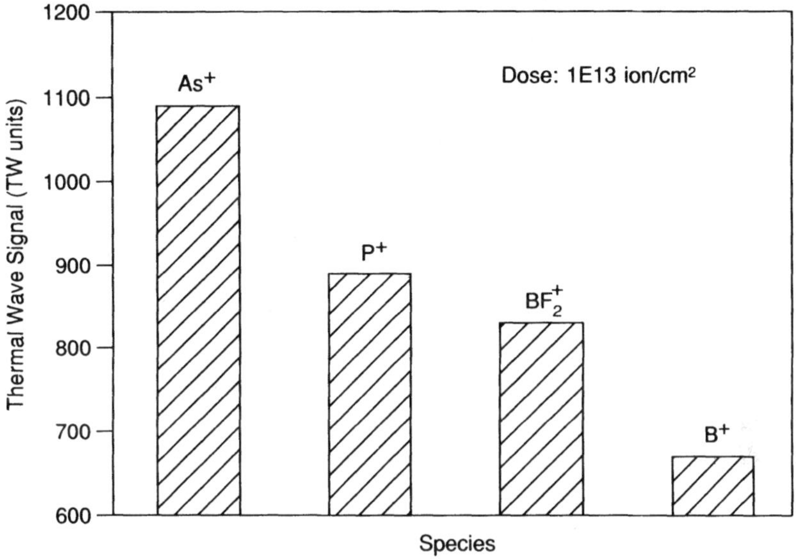

Figure 8.10 *Modulated optical reflectance signals from a set of wafers implanted with the same dose of the four species of dopant ion indicated in the figure (Rosencwaig, 1987; reproduced with permission).*

and, therefore, in the local thermoreflectance over that of unprocessed crystalline material. In effect, the 'thermal wave' produced in the sample by periodic heating will be strengthened in local amplitude but diminished in spatial extent by the ion implantation process.

The MOR technique is now used commercially for the characterization and evaluation of semiconductor wafers by the integrated circuit manufacturing industry. The main benefits of the MOR technique are its ability to measure processing damage or implant dose non-contactively and non-destructively with high spatial resolution. Figures 8.9–8.13 are examples of wafer characterization.

Figure 8.9 is an example of the detection of damage produced during the processing of silicon wafers. After polishing, a scrubbing step, with either a brush or a water jet, is used in the final cleaning cycle. A group of wafers was divided into three, one was not scrubbed and the other two were brush or jet scrubbed. Low MOR signals were obtained from un-scrubbed wafers and a large

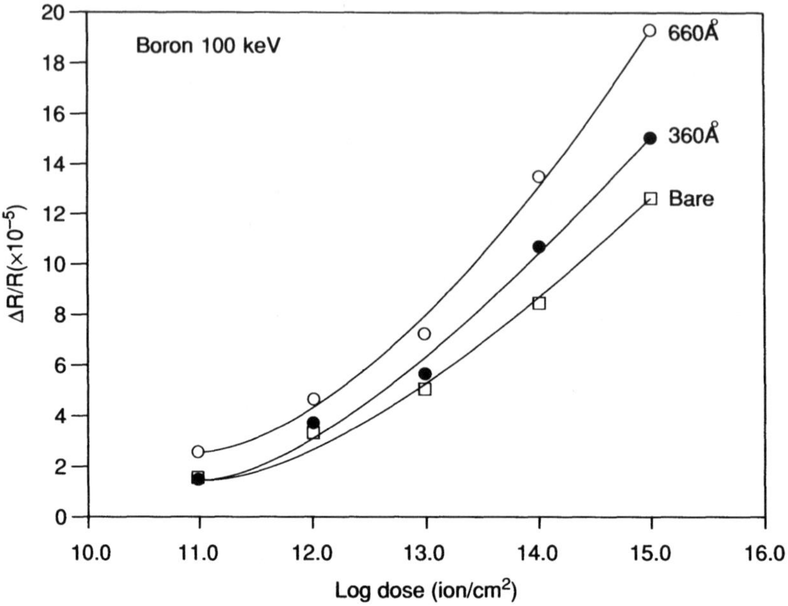

Figure 8.11 *Modulated optical reflectance signals versus dose for boron ions implanted into silicon wafers through the oxide thicknesses indicated in the figure (Rosencwaig, 1987; reproduced with permission).*

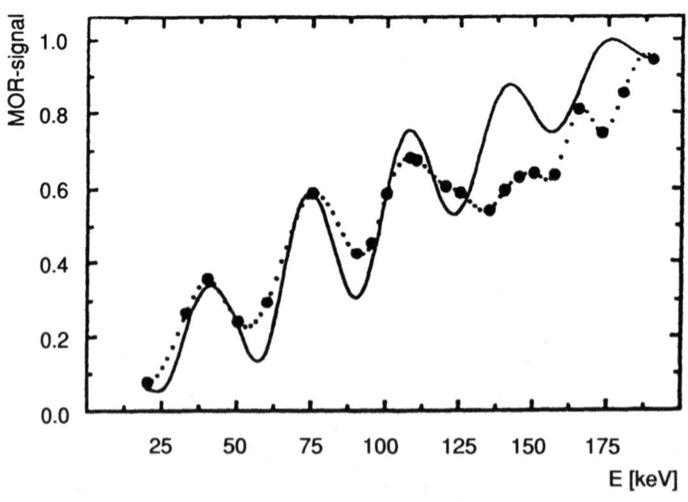

Figure 8.12 *Modulated optical reflectance signal dependence on implantation energy. The solid line is the expected theoretical variation (Wurm et al., 1988; reproduced with permission. Copyright Springer-Verlag).*

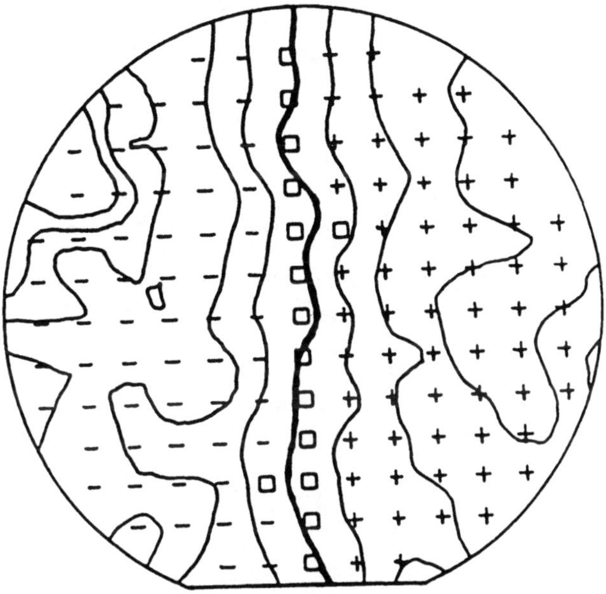

Figure 8.13 *A contour map of boron ion dose in a silicon wafer obtained by MOR. The contour interval is 0.5% (Rosencwaig, 1987; reproduced with permission).*

increase in signal was found for the others. This increased signal amplitude was attributed to surface damage which increases the surface recombination velocity.

Figure 8.10 shows the variation in the MOR signal produced by a range of implant species, indicating the possibility of detecting the dose implant species via MOR signal amplitude measurement. A major application of the MOR technique is its ability to monitor implant dose. Figure 8.11 shows the dramatic increase in the MOR signal amplitude with dose concentration, and in Fig. 8.12, with implant species energy. The increase in MOR signal with dose concentration results from the local decrease in carrier diffusivity with dose concentration. The increase in MOR signal with implant energy is caused by the accompanying increase in the extent of lattice damage. The technique is well suited to providing information about the spatial variation of dopant dose. An example of this application is shown in Fig. 8.13 which shows an MOR contour map of dopant dose in a silicon wafer. In this example

MOR reveals a smooth gradation of dose from the left to the right side of the wafer.

References

Fournier, D., Boccara, C., Skumanich, A. and Amer, N. A., 'Photothermal investigation of transport in semiconductors: Theory and experiment', *J. Appl. Phys.* **59** (1986) 787–95.

Liu, M., Suddendorf, M. B., Somekh, M. G. and Sheard, S. J., 'Modelling of photoreflectance phenomena in layered media', *Semicond. Sci. Technol.*, **8** (1993) 1630–47.

Pinto Neto, A., Vargas, H., Leite, N. F. and Miranda, L. C. M., 'Photoacoustic characterization of semiconductors: Transport properties and thermal diffusivity in GaAs and Si', *Phys. Rev. B.*, **41** (1990) 9971–79.

Rosencwaig, A., 'Thermal wave characterisation of semiconductors', in *Photoacoustic and thermal wave phenomena in semiconductors* (ed. A. Mandelis), North-Holland, New York (1987).

Wurm, S., Alpern, P., Savignac, D. and Kakoschke, R., 'Modulated optical reflectance measurements on amorphous silicon layers and detection of residual defects', *Appl. Phys. A.*, **47** (1988) 147–55.

Further Reading

Mandelis, A. (ed.) *Photoacoustic and thermal wave phenomena in semiconductors* North-Holland, New York (1987).

9

Thermal property determination by photothermal techniques

There is an undiminishing demand for reliable measurements of the basic thermal properties of the vast range of materials in use today. The magnitudes of the thermal conductivity, specific heat etc. of materials subjected to all manner of conditions are required to feed the computer models that have become the bed-rock of the modern design process. Original thermal property determinations are required for new materials and new determinations of the properties of many of the long used materials are known to be advisable. This is particularly true of thermal conductivity which can vary considerably with the particular composition and microstructure produced in a material during a specific production process.

The determination of the thermal properties has become an important well-established application of photothermal science. The 'laser flash' technique, a photothermal method, is now an internationally recognized standard method for thermal diffusivity determination. In addition, all of the other photothermal techniques that have been introduced in this book have been shown to offer ways of measuring thermal properties and/or their variation with a condition variable such as temperature. The techniques have proved attractive for thermal property measurement because they combine the non-contactive application of well defined amounts of thermal energy with methods, mostly non-contactive, for the determination of the resulting temperature changes. The temperature changes employed in a property determination can be very small because of the very high sensitivity detection techniques that have been developed in the photothermal sciences. This attribute is particularly important where measurements are required of a material whose thermal properties vary significantly with temperature. The study of phase transitions, at which thermal properties vary very rapidly with temperature, is a good example of such a situation and one in which photothermal techniques have proved successful. Another attribute is the possibility of focusing the photothermal probe and detection system to perform measurements on far smaller samples than would be needed by rival methods and to offer a means of studying the spatial variation of thermal properties across an inhomogeneous sample.

A detailed discussion of each of the thermal properties that can be obtained using photothermal techniques was given in Chapter 2. It was shown in that chapter, and in others, that a material's photothermal response depends on the two parameters: thermal diffusivity and thermal effusivity. Consequently, these are the thermal properties that can be obtained by photothermal methods rather than the primary quantities, thermal conductivity and specific heat.

9.1 Measurement of thermal diffusivity in bulk media

$$\alpha = k / \rho c \tag{9.1}$$

The thermal diffusivity is a thermal transport parameter which determines the rate of heat diffusion in a medium. Its magnitude, equation (9.1), is determined by the ratio of thermal conductivity k to thermal density, the product of specific heat c and density ρ. Table 2.1 provides an indication of the very wide range of thermal diffusivities exhibited by common materials. In general, the bulk of this variation is a reflection of the variation in the thermal conductivities of the materials rather than their specific heats or densities.

Both frequency and time-domain photothermal techniques have been used for thermal diffusivity evaluation. In each case the measurement is based on the detection of a phase or time delay in the photothermal response with respect to the applied thermal excitation. All can be regarded to be modern developments of the classic Ångström method mentioned in Chapter 1. As such, they may be classified as being non-stationary heat flow techniques. This type of technique has gained acceptance because it is less affected by heat losses, which are difficult to quantify, than a stationary heat flow technique.

By 'bulk media' we mean solid samples which are thermally thick (of thickness significantly greater than the thermal diffusion length) and for which measurements must be made assuming access is restricted to one face. For bulk solids, only those techniques which can offer a localization of heating and signal detection can be used (PTR, PTD, MOR, OBD), whilst for coatings or thin samples other photothermal methods can also be utilized.

9.1.1 The diverging-wave method

This is a simple photothermal implementation of the Ångström method. A pump laser is used to produce a modulated heating at a point on the surface of a sample and a probe is used to measure the phase-lag in surface temperature at a point some distance away, as illustrated in Fig. 9.1. Assuming a fixed separation of heating and detection points, this phase-lag depends only on the sample's thermal diffusivity if the sample is optically opaque and semi-infinite. In general, the actual separation of the points is unknown because the 'points' are diffuse spots whose profiles are determined by the focusing lenses. For this reason the technique is best used in a comparative fashion. The system is calibrated with a sample

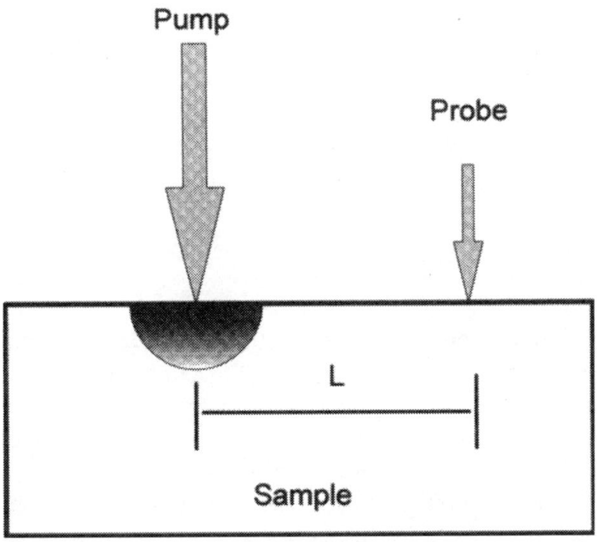

Figure 9.1 *Schematic diagram of the photothermal implementation of the Ångström method for diffusivity assessment.*

of known thermal diffusivity, after which phase-lags from unknown samples can be used to determine their thermal diffusivities.

For conventional periodic illumination, angular frequency ω, and a point separation L, the phase-lag measured on a calibration sample with known thermal diffusivity α_c would be:

$$\phi_c = L\sqrt{\left(\frac{\omega}{2\alpha_c}\right)} \tag{9.2}$$

allowing an unknown diffusivity α of a sample to be obtained from its phase-lag:

$$\phi = L\sqrt{\left(\frac{\omega}{2\alpha}\right)} \tag{9.3}$$

The MOR and PTD techniques, with pump and probe beams arranged as indicated in Fig. 9.1, have been shown to provide diffusivity measures by this method. However, the demanding experimental arrangements of these techniques, together with the requirements of a high quality sample surface and optical opaqueness, limits the range of application of this method. The very localized excitation and detection schemes used by the MOR and PTD techniques do, however, offer the possibility of making thermal property measurements on very small volume samples, typically 50 μm^3.

9.1.2 Converging-wave method

The above diverging-wave method is particularly inefficient because the probe only senses a small fraction of the thermal energy diverging from the heated point. In the converging-wave method the pump laser beam is passed through an axicon lens to produce an annular heat source at the surface and the temperature is measured at the center. The resulting temperature at the center, arising from heat converging to the center from all points on the heated annulus, has been shown to be over an order of magnitude larger than that attained using the diverging-wave configuration. This advantage provides a considerable improvement in the intrinsic signal to noise ratio of this technique.

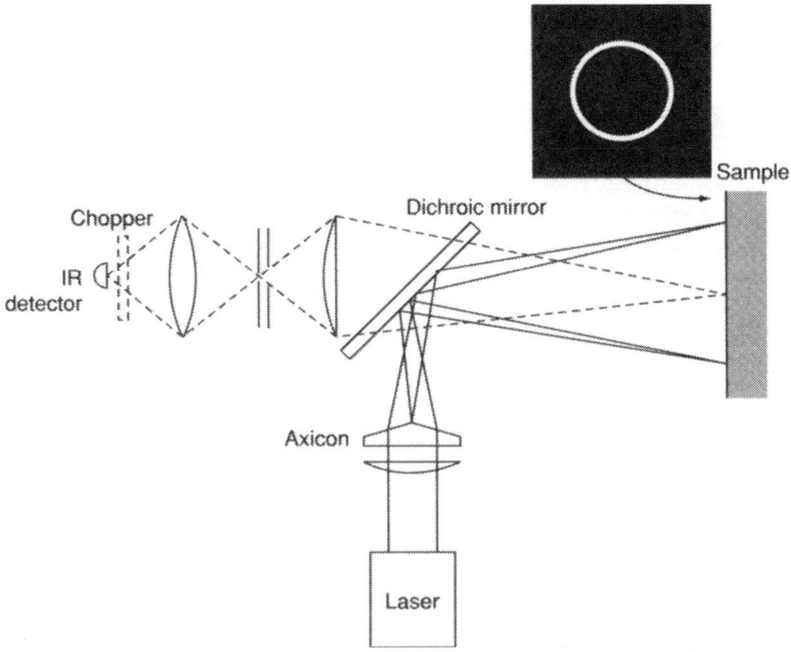

Figure 9.2 *Schematic diagram of the converging-wave apparatus for diffusivity measurement (Cielo et al., 1986; reproduced with permission).*

Figure 9.2 is a schematic diagram of the apparatus used in the converging-wave method. A pulsed laser is used to produce a heated ring on the sample surface and a focused infrared detector monitors the resulting temperature at the ring center. It has been developed as a time-domain technique in which the detector measures the variation of ring center temperature with time. For a circular heat source this temperature variation is given by the expression:

$$T = \left[E / 4\rho c(\pi\alpha t)^{p}\right]\exp(-R^{2}/4\alpha t) \tag{9.4}$$

in which R is the ring radius and p is a parameter which has a value of 3/2 for thick opaque samples and 1 for thin or translucent samples. The time for such a transient to reach half its peak value, $t_{1/2}$, is usually measured and thermal diffusivity is obtained from the expression:

$$\alpha = R^2 / mt_{1/2} \tag{9.5}$$

where $m = 13.7$ for a thick opaque sample and 10.7 for a thin or translucent sample. In transient property determinations $t_{1/2}$ is often measured in preference to the time at which the transient reaches a maximum value because the latter is often ill-defined because of the breadth of the transient peak. Typical accuracies achieved by this method are in the range ±5%.

9.1.3 OBD (mirage) zero crossing method

The basis of the OBD or mirage detection technique, which was discussed in detail in Chapter 4, is to measure the deflection of a probe laser beam passing through the air just above a surface heated at a point by a focused pump beam. At small distances above the surface the temperature distribution in the air will match that in the solid, enabling probe beam deflection measurements to be used to obtain information about the thermal distribution in the solid. Experimentally, the transverse probe beam deflection component is

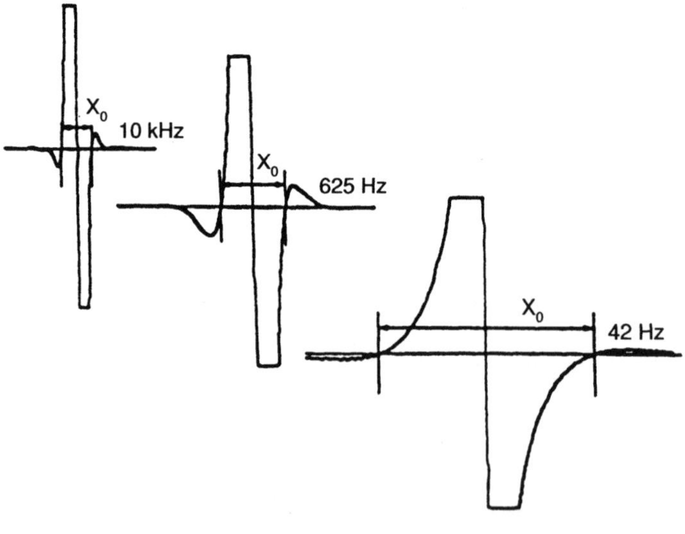

Figure 9.3 *Experimental plots of transverse mirage effect deflection as a function of separation of probe and pump beams obtained using pump modulation frequencies indicated (Kuo et al., 1986; reproduced with permission).*

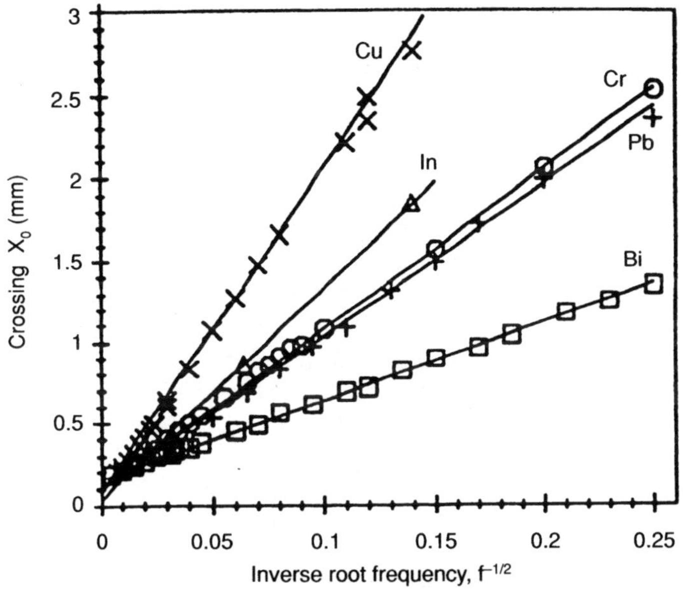

Figure 9.4 *Experimental zero crossing distances for pure elements indicated plotted as a function of root modulation frequency (Kuo et al., 1986; reproduced with permission).*

monitored as the probe beam is scanned to vary its distance x away from the heated spot, keeping its height above the surface constant. The 'zero crossing' distance x_0, corresponding to the probe signal falling to zero, is recorded. At this distance from the heated spot the thermal wave phase-lag is + or -90° on one or the other side of the spot. A set of probe beam deflection traces obtained at three different frequencies is shown in Fig. 9.3. The distance x_0 corresponds to a quarter of the thermal wavelength in the solid and is related to the modulation frequency f by:

$$x_0 = \sqrt{\gamma\alpha / f} \tag{9.6}$$

γ has a value of 1.44 for opaque thermally thick media and 1.1 for translucent or thermally thin media. Thermal diffusivity can be obtained from the gradient of a graph of zero crossing values plotted against inverse root frequency, examples are shown in Fig. 9.4. The linearity in the above relationship fails when the thermal diffusivity of the sample is much lower than that of the adjacent air or fluid. In this case no zero crossings in the probe signal component are expected. The OBD technique can, however, still be used for low diffusivity materials through a more complete data fitting of the OBD deflection components.

With this zero crossing method, an accurate determination of the thermal diffusivity is possible due to low measurement errors in the two parameters, the frequency and zero crossing position. Access to only one face of the sample is required and very small sample volumes are required for the measurements, typically ~1 mm^3. Measurements of the thermal diffusivity in different directions in the plane of the sample can also be made to obtain the anisotropy in thermal diffusivity. A disadvantage of the OBD method is the requirement of a plane sample surface which does not obstruct the skimming probe beam. Under optimum conditions thermal diffusivity values can be obtained with a 5% measurement error.

9.1.4 Transient line heating method

With the line heating method, the cooling transient is monitored in a direction perpendicular to the line heat source, at a distance x from it. The expression for the decaying temperature profile, following pulse excitation, takes the form:

$$T(x,t) = \frac{\exp(-\upsilon t)\exp\left(-x^2 / 4\alpha t\right)}{2(\pi\alpha t)^{1/2}} \tag{9.7}$$

where υ is a convective heat loss parameter. It can be seen in this equation that the heat loss term does not alter the shape of the temperature profile. A plot of the half-width versus root time can be used to obtain the thermal diffusivity of the sample. Thermal diffusivities of metals and polymer composite materials have been obtained by this method. A combination of a pulsed laser and a fast linear array infrared sensor is necessary for the determination of high diffusivities by this method.

9.1.5 Transient thermal grating method

The measurement of the decay of a transient thermal grating (TTG) is another quite different technique which is being used for thermal

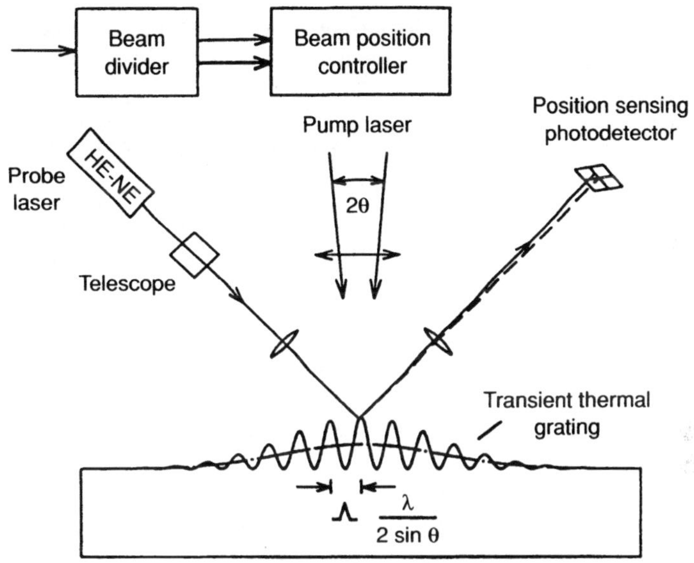

Figure 9.5 *Schematic diagram of the experimental set-up used for the generation and measurement of transient thermal gratings (Jauregui and Matthias, 1982; reproduced with permission).*

diffusivity measurements in bulk and thin film materials. A TTG is formed on a sample surface by the interference pattern of two coherent laser beam pulses. Photothermal processes convert the optical interference pattern into a periodic array of heated and unheated material. This 'thermal grating' decays at a rate determined by the grating spacing and the thermal diffusivity of the material. Thermal expansion caused by the TTG produces a periodic surface distortion which can be monitored by a probe laser. Figure 9.5 is a schematic diagram of the experimental arrangement. The periodicity of the TTG is determined by the pulse laser wavelength and the angle between the two interfering laser beams.

A three-dimensional heat flow analysis shows that the effective decay rate of the TTG is composed of the decoupled decay rates for the three directions:

$$\tau_{\text{eff}}^{-1} = \tfrac{1}{2}\tau_x^{-1} + \tau_y^{-1} + \tau_z^{-1} \tag{9.8}$$

Here, the grating periodicity is along the y-axis and its wave vector

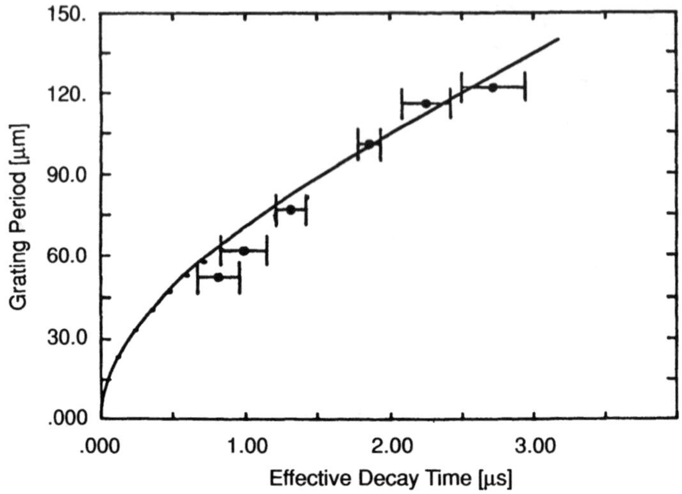

Figure 9.6 *Experimental transient thermal grating data obtained from a polished copper surface. The fitted line was obtained using equation (9.9) (Jauregui and Matthias, 1992; reproduced with permission).*

$q = 2\pi/$ is connected with τ_y^{-1} by the relation $\tau_y^{-1} = q^2\alpha$, where α is the thermal diffusivity. The decay rate along the x-axis is determined by the pulse laser beam profile and τ_z^{-1} is the vertical heat diffusion rate.

Thermal diffusivity can be obtained from measurements of the effective decay rates of gratings having several different grating periods via the equation:

$$\tau_{\text{eff}}^{-1} = \left[\frac{2}{X_0^2} + \frac{4\pi^2}{\lambda^2}\right]\alpha + \tau_z^{-1} \tag{9.9}$$

in which X_0 is the extension of the grating in the x direction, assuming a Gaussian beam profile. An example of TTG measurements of a copper sample fitted to equation (9.9) are shown in Fig. 9.6.

The TTG technique has been shown to have the capability of determining the thermal diffusivity of opaque solids and thin films with an accuracy that is limited by thermal grating spacing measurement precision. It has a potential for anisotropic thermal diffusivity measurement (in plane and normal to the sample surface) and the thermal characterization of submicron thin films, to be discussed below.

9.2 Thermal diffusivity measurement of thin samples and coatings – periodic heating methods

The thermal diffusivity of an opaque thermally thin sample or coating can be determined in several ways. Frequency-domain techniques may be divided into three types:

1. phase lag;
2. thermal wave interference;
3. dual face heating.

9.2.1 Phase lag

This is a particularly simple and direct method in which one face of a thin sample is heated and the signal reaching the opposite face is monitored. The heating is produced by conventional periodically modulated optical illumination and the signal obtained on the opposite face of a sample of thickness L is given by a general formula of the type:

$$S = S(\omega)\exp(-\sigma L + \mathrm{j}\omega t) \tag{9.10}$$

The phase variation of the signal is analyzed by varying either modulation frequency or sample thickness

$$\varphi = L\sqrt{\frac{\omega}{2\alpha}} + \varphi(f) \tag{9.11}$$

When phase versus frequency analysis is used, a normalization process is necessary to eliminate the system phase shift, $\phi(f)$. Reference measurements are made on a thermally thick and optically opaque sample. The normalized phase is then obtained by a simple subtraction of the reference sample phase from that measured at the same frequency on the thin test sample. If thin samples of several thicknesses are available, then a normalization process is unnecessary. A subtraction of the phase data obtained from different thickness samples eliminates $\varphi(f)$. The phase difference is now simply given by:

$$\Delta\varphi = \sqrt{\frac{\omega}{2\alpha}}(L_2 - L_1) \tag{9.12}$$

where L_1 and L_2 are the two sample thicknesses.

There are limits of applicability for the above equation. The assumptions made in the derivation of the equation are: that the heat flow in the sample is uni-dimensional; the sample is optically opaque; its thermal properties are homogeneous and that the transmitted thermal wave is the primary contribution to the detected signal. The latter is particularly important in photoacoustic measurements where sample vibration due to periodic thermal expansion may result in an additional photothermal signal. This mechanical vibrational signal is strong from samples with high thermal expansion coefficients, such as polymers. For these materials one of the other photothermal detection techniques should be used.

9.2.2 Thermal wave interference

The thermal wave interference concepts which were introduced in Chapter 2 and discussed as a means of coating thickness determination in Chapter 7 also provide the basis for a thermal diffusivity measurement technique that is particularly suitable for the assessment of coating material. The coating surface is heated periodically and the resulting periodic thermal response of the same surface is monitored. Thermal wave interference in a coating produces a phase change, with respect to an uncoated surface, equation (2.40):

$$\varphi = -\tan^{-1}\left[\frac{2R\exp(-x)\sin x}{1 - R^2\exp(-2x)}\right] \tag{9.13}$$

where $x=2L/\mu$, μ is the thermal diffusion length of the coating material and R is the thermal wave reflection coefficient at the coating–substrate interface. Normalization must again be used to remove system phase shifts. A non-linear least squares fitting algorithm can then be used to determine μ and R for a known coating thickness L from a set of phase versus frequency measurements. The coating thermal diffusivity is obtained from the thermal diffusion length, $\mu = \sqrt{(2\alpha/\omega)}$. From the definition of R, Section 2.4, the thermal effusivity of the coating can also be obtained if that of the substrate material is known, see Section 9.4. The analysis again makes the assumptions of optical opaqueness; coating homogeneity and uni-dimensional heat flow. The latter assumption necessitates the diameter of the heating spot being considerably larger than the thermal diffusion length. Care is needed to avoid violating this condition at low frequencies. Thermal diffusivities of translucent or transparent coatings can also be obtained by making use of the more complex expressions for thermal wave interference given in Section 3.3. Most thermal wave interference measurements of coatings have been made using infrared detection. The technique has been demonstrated to have the potential for high quality thermal property measurements, limited in precision, in many cases, by uncertainties or variations in coating thickness. It is a surprising but common problem that the practical limit on the precision of many of the photothermal techniques is the precision of sample thickness measurement.

9.2.3 Dual face heating

When both faces of a thin sample are accessible, thermal diffusivity can be obtained from a PA method in which signal and phase differences for front and rear sample face heating are measured. The phase difference between the signals is given by the expression:

$$\tan(\Delta\phi) = \tanh(L/\mu)\tan(L/\mu) \qquad (9.14)$$

The thermal diffusivities of a range of opaque thin samples of metals and semiconductors have been determined accurately by this simple dual face heating PA method.

9.3 Thermal diffusivity measurement of thin samples and coatings – transient response laser flash method

The thermal transient response of a sample following a heating pulse, usually provided by a pulsed laser, can be used for diffusivity measurement. For an opaque sample, a surface heating pulse results in a rear face thermal transient which varies as:

$$T(L,t) = T_0\left[1 + 2\sum_{n=1}^{\infty}(-1)^n \exp(-n^2\pi^2\alpha t/L^2)\right] \qquad (9.15)$$

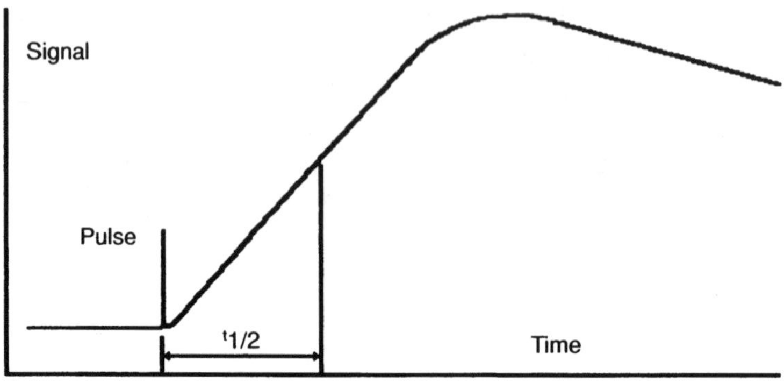

Figure 9.7 *Schematic diagram of a laser flash transient.*

where T_0 is the maximum sample temperature rise and L is its thickness. Thus thermal diffusivity can be obtained from measurements of this rear face transient. This method, first proposed by Parker *et al.* (1961), has become an international standard thermal diffusivity measurement technique. Several algorithms have been presented in the literature for the evaluation of thermal diffusivity. The most commonly used method is to utilize $t_{1/2}$, the time at which the transient reaches half its peak value. As mentioned before, this time is generally better defined than the time for the transient to peak, see Fig. 9.7. The diffusivity is obtained from the simple expression:

$$\alpha = \frac{0.139L^2}{t_{1/2}} \tag{9.16}$$

Figure 9.8 is a schematic diagram of a laser flash system. Commonly used lasers are ruby or Nd doped glass. These types of laser can provide sufficient pulse energy to produce a temperature rise of over a degree in most materials. Their pulse lengths are below 1 ms and beam diameters are large, about 1 cm, to ensure uni-dimensional heat flow. The sample is usually a disk with a thickness dependent on the diffusivity of the material concerned. The sample thickness is chosen to ensure that pulse length is less than about 1% of $t_{1/2}$, to minimize errors associated with the finite laser pulse length. The sample may be housed in an experimental chamber to enable measurements to be performed at high or low temperature.

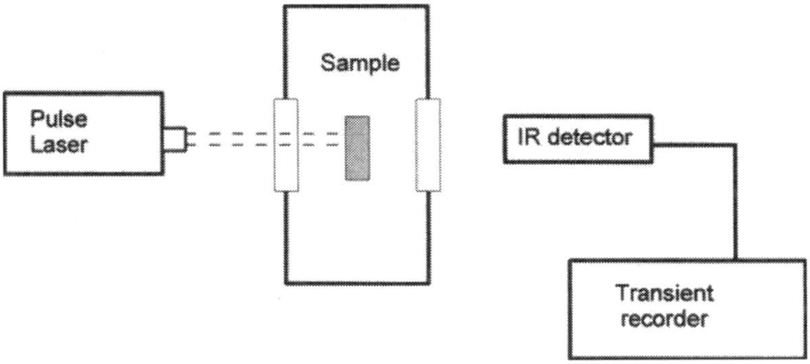

Figure 9.8 *Schematic diagram of a laser flash diffusivity measurement apparatus.*

A semiconductor infrared detector, e.g. InSb, is used to record the thermal transient at the unheated face of the sample.

The laser flash technique has been applied to a very wide range of materials with diffusivities from 10^{-7} to 10^{-3} m^2 s^{-1} at temperatures from 100 to 3300K. The accuracy of measurement depends critically on the preparation of a homogeneous parallel sided sample with polished flat faces and on the measurement of the sample thickness. The measurement of $t_{1/2}$ is affected by the signal-to-noise ratio achieved in recording the transient and by the real finite laser pulse length. For highest accuracy, detailed curve fitting methods, utilizing all the data in a transient, are used instead of relying on the $t_{1/2}$ measurement alone. In an ideal case a measurement uncertainty of between 1% and 2% can be achieved.

If the sample is a coating on a substrate, the thermal diffusivity of the coating material can still be obtained but now both the heating and detection must take place at the sample surface. For an opaque coating heated by a plane pulsed heat source the surface temperature decays as (Section 5.3, equation (5.15)):

$$T(0,t) = T_0\left[1 + 2\sum_{n=1}^{\infty} R^n \exp(-n^2L^2/\alpha t)\right] \tag{9.17}$$

where R is the coating/substrate thermal wave reflection coefficient. A data fitting procedure can be applied to determine the thermal diffusivity of the sample assuming that the coating thickness is known.

These transient response techniques have become widely adopted because they have been shown to be rapid, accurate and reliable. They are also remote, non-contactive techniques that can be used to measure samples in extreme environments, such as at high temperatures in a vacuum furnace. During the last few decades the laser flash technique has become the major means of generating thermal diffusivity data.

9.4 Thermal effusivity determination

The thermal effusivity e, of opaque solid material can be determined from both frequency and time-domain measurements of semi-infinite samples. For example, with measurements using photoacoustic detection the signal varies as:

$$S(f) = \frac{S_0[1 - R(\lambda)]}{e} f^{-3/2} \tag{9.18}$$

where S_0 is a constant parameter independent of the thermal properties of the material. If two measurements are made: one on a sample with known thermal properties and another on the test sample then the ratio of the photoacoustic signals eliminates S_0 and the frequency. This ratio provides the thermal effusivity of the unknown sample e_s in terms of that of the known sample e_r:

$$e_s = \frac{[1 - R(\lambda)]_s}{[1 - R(\lambda)]_r} \frac{S_r}{S_s} e_r \tag{9.19}$$

The dependence on the optical reflectivities can be eliminated by reflectivity measurements on the two samples. Although similar expressions can be derived for other photothermal detection techniques, an explicit expression cannot be obtained for the thermal effusivity because of the dependence of the measured signal on optical properties. For example, emissivity affects PTR measurements, $n\, dn/dT$ affects OBD measurements and $(1/R).(dR/dT)$ affects MOR measurements.

A novel approach to thermal effusivity measurement, which can be applied to all types of material, is to use the PPE technique with the PPE sensor at the sample surface. This is the so called inverse PPE method. Under certain experimental conditions the PPE signal depends only on the thermal effusivity of the substrate material. If the PPE sensor is thermally thin, optically opaque and operates in the current mode ($\omega\tau_e < 1$), the PPE voltage is given by the expression (Section 4.9):

$$U = B(f)e^{-1/2} \tag{9.20}$$

where *B(f)* is a constant depending only on the optical and thermal properties of the PPE sensor. After calibration with a sample of known thermal properties, the constant *B(f)* can be eliminated, enabling the thermal effusivity of an unknown sample to be evaluated. This method has been applied to metals, liquids and biological materials.

The thermal effusivity of a coating can be obtained by thermal wave interference, as mentioned in Section 9.2.2, from the determination of the coating/substrate thermal wave reflection coefficient. Thermal wave phase data obtained from coatings of differing thicknesses, or at different frequencies from a coating of known thickness, can be fitted by equation (9.13) to obtain the thermal wave reflection coefficient R, equation (2.35):

$$R = \frac{1 - e_s / e_c}{1 + e_s / e_c} \tag{9.21}$$

from which the coating material effusivity e_c can be obtained provided that of the substrate material e_s is known. Equation (9.13) is valid for opaque coating material. Thermal wave interference in translucent and transparent material leads to different expressions, as discussed in Chapter 3.

9.5 Thermal conductivity determination

The thermal conductivities of most materials are given by the expression:

$$k = \frac{1}{3}\rho c_p v l \tag{9.22}$$

where ρ is the density, c_p the specific heat, v the mean carrier (electron or phonon or both) velocity and l the carrier mean free path.

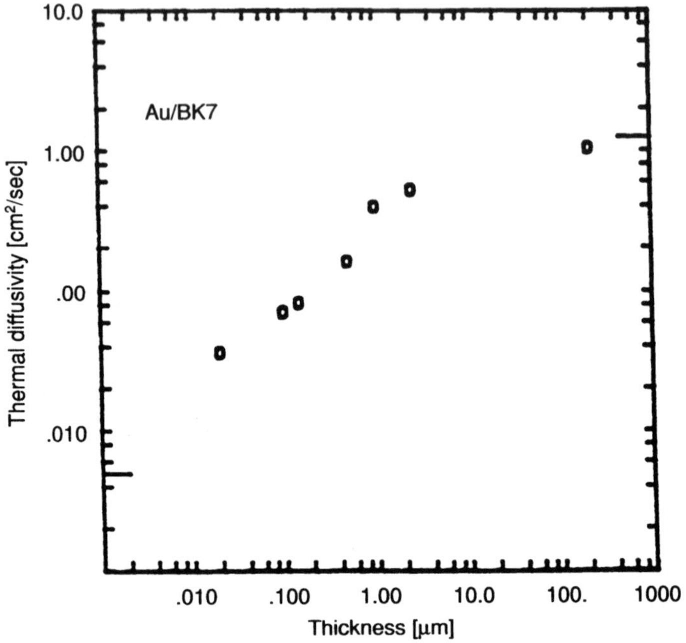

Figure 9.9 *Thickness dependence of the thermal diffusivity of gold films on BK7 glass (Jauregui et al., 1992; reproduced with permission).*

Thus thermal conductivity depends on the specific heat, which is affected by compositional variables, and on details of microstructure and material imperfections (impurities, grain boundaries, thickness etc.) which modify the electronic and phonon velocity and mean free path.

Thermal conductivity values can be obtained from the measurements of the thermal diffusivity and effusivity by applying the relationship:

$$k = e\sqrt{\alpha} \quad \text{or} \quad k = \alpha \rho c \tag{9.23}$$

Hence photothermal measurements provide the means of obtaining α, e and k values for bulk or thin samples. No photothermal methods can give the thermal conductivity of a sample directly since this parameter is always coupled with the heat capacity.

The thermal conductivity/diffusivity of thin films deserves a special mention, particularly when the film thickness is in the micrometer range. In this thickness range the thermal conductivity/diffusivity can be a function of the film thickness as the mean free paths of the carriers may be comparable to the film thickness and because surface disorder and damage may amount to a significant proportion of the film thickness. Photothermal measurements have shown that thin film thermal conductivities/diffusivities can be considerably lower than obtained for bulk material. This observation has important implications for the measurement of film thickness by photothermal methods. Changes observed in signal and phase will have contributions due to changes in both film thickness and thermal conductivity/diffusivity. Figure 9.9 shows the variation in the thermal diffusivity of gold films with thicknesses in the range 0.1 to 200 μm. The gold films were deposited on glass and the measurements were made using the TTG technique. There is a strong decrease in the thermal diffusivity of the gold film below 1 μm which is attributed to both film geometry and structural disorder, both of which impede the lateral heat flow which is measured by the TTG technique.

9.6 Specific heat determination

Specific heat capacity values can also be obtained from photothermal measurements using the relationship:

$$\rho c = e / \sqrt{\alpha} \tag{9.24}$$

In addition to this relationship, under certain experimental conditions the photothermal signal variation can be inversely proportional to the sample's specific heat. This is the case with the PA, PTR, OBD and MOR detection techniques when the sample is optically translucent. This condition is usually met at high modulation frequencies. With the PZT-PA method the optical transparency condition is not necessary for signal dependence on specific heat. For the PPE method, an inverse dependence of the PPE signal on specific heat can be obtained in optically opaque thin films at low modulation frequencies:

$$V(\omega) \propto \frac{f(T)}{\omega^n c_s(T)} \tag{9.25}$$

where $n = 2/3$ for a thermally thin and unity for a thermally thick PPE transducer. $f(T)$ is a function which describes the temperature dependence of the PPE cell. A direct evaluation of the specific heat capacity is therefore possible through an elimination of the function $f(T)$ using a reference sample

$$c_s(T) = V_r(T)c_r(T) / V_s(T) \tag{9.26}$$

The PPE method also offers an increased signal-to-noise ratio with respect to PA technique as measurements are made at low frequencies and thus high signal amplitudes.

References

Cielo, P., Utracki, L. A. and Lamontagne, M., 'Thermal-diffusivity measurements by the converging-thermal-wave technique', *Can. J. Phys.*, **64** (1986) 1172–7.

Kuo, P.K., Lin, M. J., Reyes, C. B., Favro, L. D., Thomas, R. L., Kim, D. S., Zhang, S. Y., Inglehart, L. J., Fournier, D., Boccara, A. C. and Yacoubi, N., 'Mirage-effect measurements of thermal diffusivity', *Can. J. Phys.*, **64** (1986) 1165–71.

Jauregui, J. and Matthias, E., 'Thermal diffusivities measured by photodisplacement detection of transient thermal gratings', *Appl. Phys. A.*, **45** (1992) 35–39.

Jauregui, J., Wu, Z. L., Schafer, D. and Matthias, E., 'Thermal diffusivities of thin gold films', in *Springer Series in Optical Sciences Vol 69: Photoacoustic and photothermal phenomena III* (ed. D. Bicanic) (1992) 682–7.

Parker, W. J., Jenkins, R. J., Butler, C. P. and Abbot, G. L., 'Thermal diffusivity measurement using the flash technique', *J. Appl. Phys.*, **32** (1961) 1679–84.

Further reading

Bertolotti, M., Li Voti, R., Liakhou, G. and Sibilia, C., 'On the photodeflection method applied to low thermal diffusivity measurements', *Rev. Sci. Instrum.* **64** (1993) 1576–83.

Dadarlat, D., Chritoc, M., Nematu, C., Candea, R. M. and Bicanic, D., 'Inverse photopyroelectric detection method', *Phys. Stat. Sol. K.*, **121** (1990) 231–34.

Gembarovic, J., Vozar, L. and Majernik, V., 'Using the least square method for data reduction in the laser flash method', *Int. J. Heat and Mass Trans.*, **33** (1990) 1563–65.

Kim, S. W. and Taylor, R. E., 'Estimation of thermophysical properties of a film coated on a substrate using pulsed transient analysis', *Int. J. Thermophys.*, **14** (1993) 135–47.

Lachaine, A. and Poulet, P., 'Photoacoustic measurement of thermal properties of a thin polyester film', *Appl. Phys. Lett.*, **45** (1984) 953–5.

Maglic, K. D. and Taylor, R. E., 'The apparatus for thermal diffusivity measurement by the laser pulse method', in *Compendium of thermophysical property measurement methods 2* (eds K. D. Maglic, A. Cezairliyan and V. E. Peletsky), Plenum Press (1992).

Pessoa, O., Cesar, C. L., Patel, N. A. and Vargas, H., 'Two beam photoacoustic phase measurement on the diffusivity of solids', *J. Appl. Phys.*, **59** (1986) 1316–18.

Quelin, X., Perrin, B., Louis, G. and Peretti, P., 'Three dimensional thermal conductivity-tensor measurement of a polymer crystal by the photothermal probe beam deflection method', *Phys. Rev. B.*, **48** (1993) 3677–92.

Salazar, A., Sanchez-Lavaga, A. and Fernadez, J., 'Theory of thermal diffusivity determination by the "mirage effect" techniques in solids', *J. Appl. Phys.*, **65** (1989) 4150–6.

Salazar, A., Sanchez-Lavaga, A. and Fernadez, J., 'Thermal diffusivity measurement on solids using collinear mirage detection', *J. Appl. Phys.*, **74** (1993) 1539–47.

10

Applications of photothermal techniques in gases and liquids

10.1 Introduction

The main areas of application of the photothermal effect in gases and liquids are for:

1. spectral studies (spectroscopy, absorption coefficient determination);
2. environmental pollution studies (monitoring of trace concentrations of gases, elements and compounds);
3. combustion diagnostics;
4. fluid flow measurements;
5. thermal conductivity and diffusivity measurements.

For these applications, both the photoacoustic and photothermal beam deflection methods have been utilized. In the following sections we present the basic theory and experimental techniques employed in this area. This is then followed by a range of examples of reported experimental results.

10.2 Photoacoustic signal generation in gases

Photoacoustic signal generation is essentially a two stage process: heat production and the generation of acoustic waves. An absorbed photon may excite a molecule from its ground state energy E_0 to an excited state energy E_1

$$\Delta E = E_1 - E_0 = h\nu \tag{10.1}$$

The molecule can loose this excitation energy by a number of mechanisms:

1. reradiation of a photon;
2. the initiation of a photochemical change such as a bond alteration;
3. collision with other molecules and excitation to a higher energy state;
4. collision with other molecules and transference of energy to translational energy (vibrational–translation, vibrational–vibrational translational–translational or rotational to rotational etc.).

Heat production in a gas occurs by the non-radiative relaxation process (4), although heat may also eventually result from the down-conversion of the excitations produced by processes (2) and (3). Assuming a one component gas and in the absence of optical saturation with infrared illumination, the heat generation rate is given by the expression:

$$H = N\sigma I_0 \qquad (10.2)$$

at normal atmospheric pressure and temperature. N is the total molecular number density and σ is the absorption cross-section. Where the illumination source I_0 is modulated, the heat generation is in phase with the illumination.

In chemical trace gas analysis or air pollution studies, a multi-component gas mixture is generally encountered. In this case the above equation is modified to:

$$H = I_0 N_{tot} \sum_{j=1}^{n} c_j \sigma_j \qquad (10.3)$$

where c_j and σ_j are the concentration and absorption cross-sections of the species j.

The final expression for the PA signal depends on the type of PA cell used, resonant or non-resonant, and the geometry of the cell determining the dominant acoustic modes. For a commonly used resonant PA cell, the PA signal for trace gas studies can be written in the form:

$$S = CPN_{\mathrm{tot}} \sum_{j=1}^{n} c_j \sigma_j \qquad (10.4)$$

where P represents the average laser power related to intensity I_0 and C is a cell constant.

10.3 Photoacoustic gas cells

A photoacoustic system adapted for the study of gases is shown schematically in Fig. 10.1. The modulated laser beam passes through a photoacoustic cell containing the gas under investigation. The major design criteria for a PA cell suitable for the study of gases include: the consideration of the cell material; cell volume; modulation frequency and the type of measurements to be made (closed cell or flowing gas cell). Some typical examples of PA cells developed for the evaluation of gases are shown in Fig. 10.2. There are essentially two types of PA cells, resonant and non-resonant. The latter offer excellent performance at low frequencies and employ small cell volumes. The resonant cell has the advantages of operating at high modulation frequencies ($1/f$ noise lower) and of providing higher signal amplitudes due to the high Q factor (> 100) of the cell. For very low optical absorption gases, reflecting mirrors can be used to make the light beam travel several times through the cell to increase the fraction of light energy absorbed by the gas sample to increase signal amplitude.

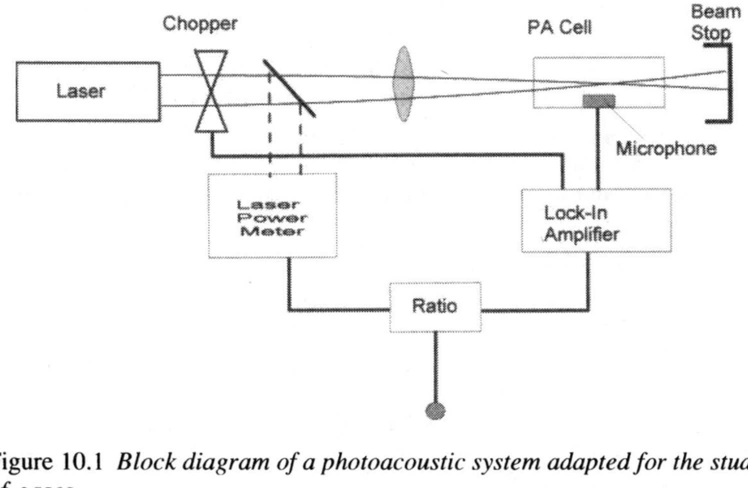

Figure 10.1 *Block diagram of a photoacoustic system adapted for the study of gases.*

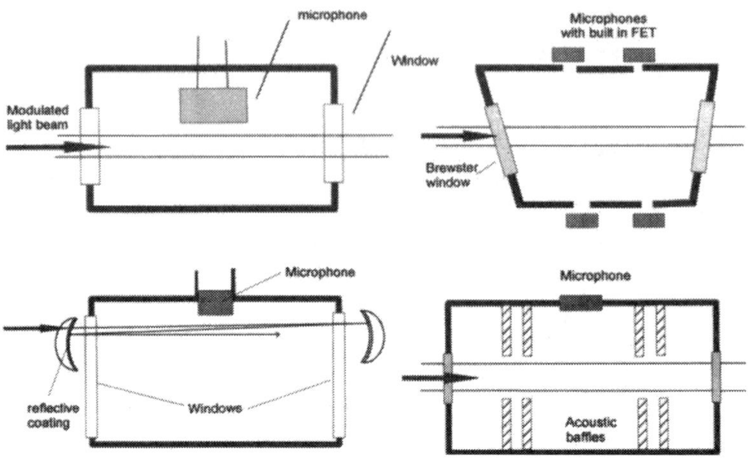

Figure 10.2 *Schematic diagrams of photoacoustic cells designed for use with gases.*

The main noise sources in PA gas cells arise from signal generation by light absorption at cell windows and walls. In trace gas detection, absorption by other gas molecules in the cell can contribute to the overall PA signal and it may be necessary to prevent these gases from entering the PA cell or to operate at wavelengths well removed from their absorption peaks. The cell and detection scheme can be designed to reduce these coherent noise sources. PA noise signals resulting from absorption in cell windows can be reduced by the use of acoustic baffles placed close to the cell windows. Highly polished cell walls can reduce PA noise caused by optical absorption at the cell walls. However, owing to their coherency with the required PA signal from absorption in the gas, it is difficult to eliminate completely these types of noise signal in conventional periodic heating PA systems with lock-in signal detection. This has led to interest in the pulsed heating mode and box-car detection of the PA signal. The difference in the thermal diffusivity of the window material and the gas sample results in PA signals that are sufficiently separated in time to allow discrimination of the cell noise components from the required gaseous component. An alternative solution to the inherent noise problems in PA cells is to use the photothermal refraction based detection techniques described below.

10.4 Photoacoustic spectroscopy of gases

The PA method offers several advantages over conventional techniques for spectroscopic studies of gases:

1. a wide dynamic range, typically over five decades at least, enabling the determination of trace concentrations of gases from ultralow to high;

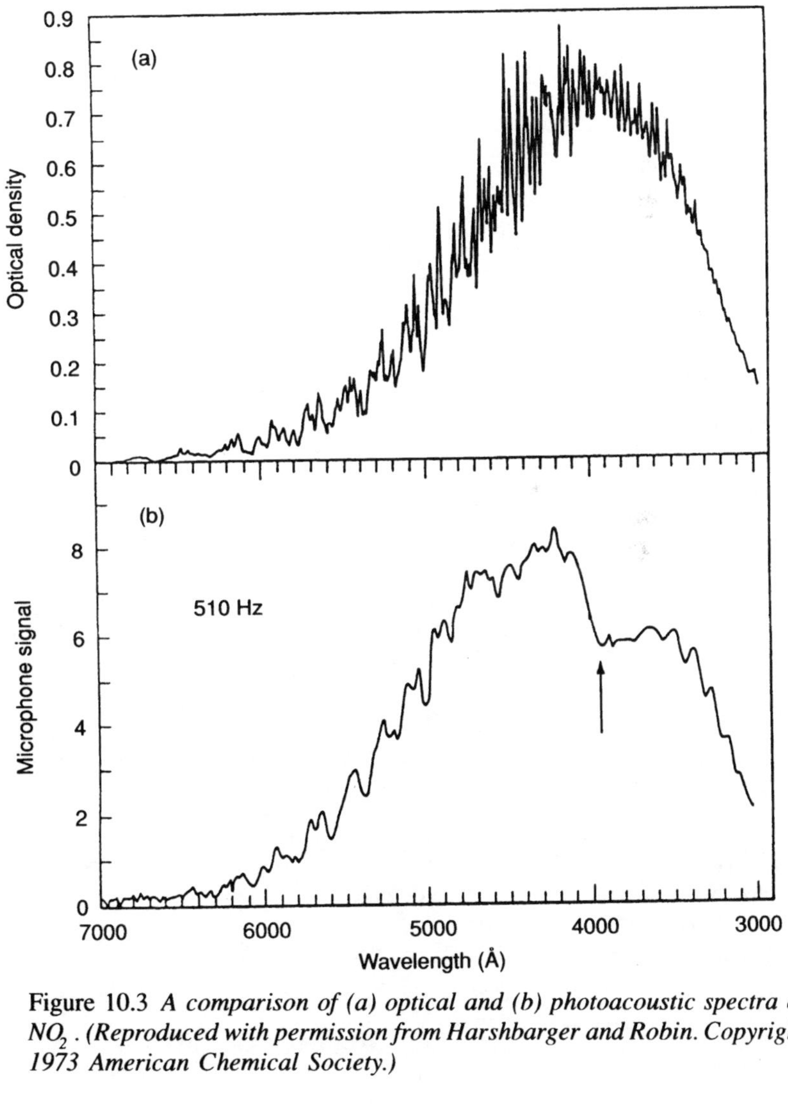

Figure 10.3 *A comparison of (a) optical and (b) photoacoustic spectra of* NO_2. *(Reproduced with permission from Harshbarger and Robin. Copyright 1973 American Chemical Society.)*

2. a high sensitivity, allowing optical absorption measurements of gases at trace concentration levels ($\beta = 10^{-8}$ cm^{-1});
3. good detection selectivity achievable by matching the excitation source wavelength to the required absorbing species;
4. a simple experimental set-up which provides wavelength independent responsivity and room temperature operation.

An example of a PA and an optical absorption spectrum of a gas sample is shown in Fig. 10.3. The gas is NO_2 and the photoacoustic spectrum was obtained by a double beam spectrometer with carbon black in the reference channel. This spectrum shows absorption features related to electronic relaxation processes and a signal fall at 400 nm associated with the photochemical dissociation: $NO_2 \rightarrow NO + O$. The latter effect was not evident in the conventional optical absorption spectrum, showing that photo-induced chemical changes can also be studied by PA. This result is not unexpected as the PA signal is generated from changes in the heat energy of the excited sample.

10.4.1 Pollution gases

The pollution of the atmosphere from a range of sources has become a major concern throughout the world. A wide range of pollutant gases arising from domestic and industrial sources need to be regulated and monitored. Typical environmental emission sources are summarized in Table 10.1.

The photothermal techniques, in particular photoacoustics, have been extensively developed and portable systems have been utilized in the field to monitor pollution gases. The two most commonly

Table 10.1 Environmental emission sources

Source/Location	Species
Ambient air	NO, C_2H_2, NH_3, H_2O, CO_2
Stratosphere	NO
Automobile exhaust	NO, hydrocarbons
Spoiled beef	NH_3
Fumigation chamber	NH_3
Fruit storage chamber	C_2H_2
Greenhouses	C_2H_2
Stack gases	SO_2, NO_x, NH_2, CO, CO_2, H_2O, O_3

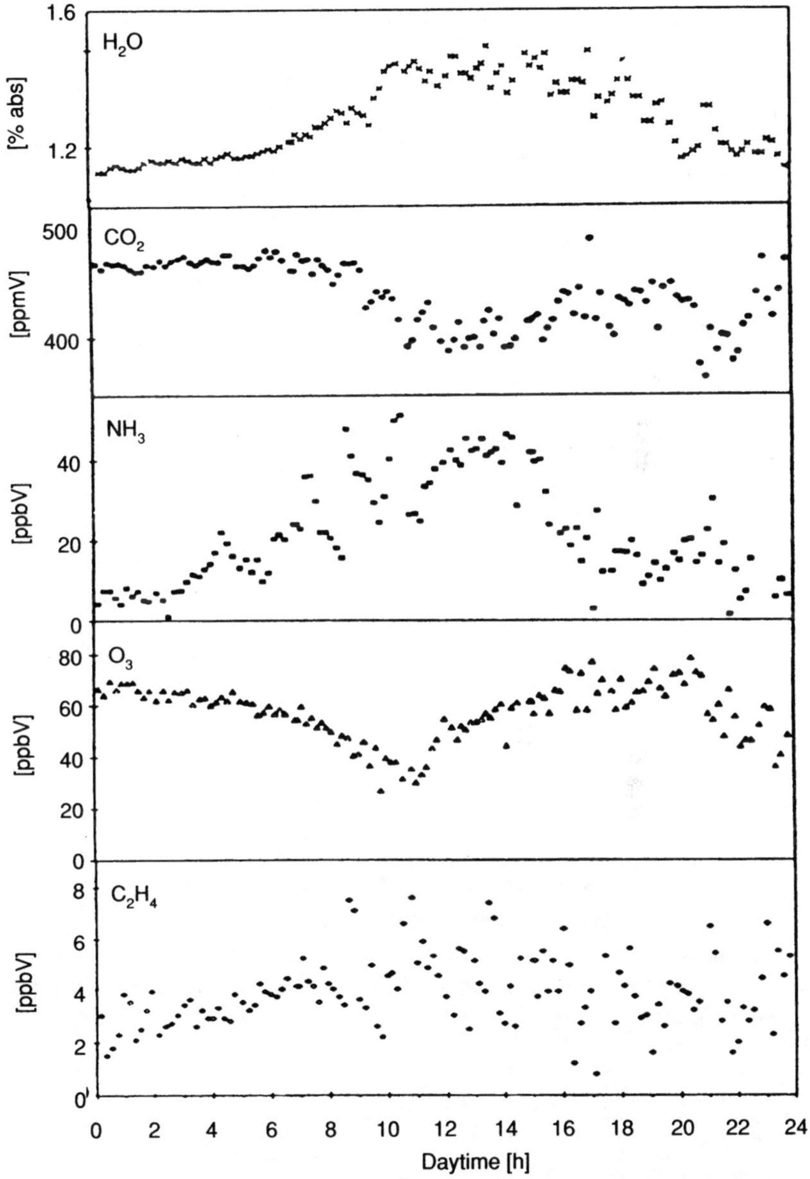

Figure 10.4 *The temporal concentrations of H_2O vapour. CO_2, NH_3, O_3 and C_2H_4 in rural air, obtained using a mobile CO_2 laser PA system at Morschach, central Switzerland (Sigrist, 1994; reproduced with permission).*

used sources in environmental studies are CO and CO_2 lasers. Figure 10.4 is an example of PA measurements of the concentrations of H_2O, CO_2, NH_3, O_3 and C_2H_4 over a period of 24 h obtained in a field trial. The variation of the various gas concentrations was explained by taking into consideration the local climatic conditions (wind direction, rain, temperature) and the location of the measurement site (whether near road, factory etc.). A detailed discussion of the interpretation of this data is given by Sigrist (1994).

10.5 Photothermal refractive index modulation techniques

Periodic photothermal heating produces modulations in the refractive index of a medium because of the temperature dependence of the refractive index n:

$$n(x,y,t) = n_0 + \frac{\partial n}{\partial T} T(x,y,t) \tag{10.5}$$

The temperature distribution $T(x, y, t)$, for a weakly absorbing medium, pulse heated by a Gaussian profile laser beam, is given by:

$$T(x,y,t) = \frac{2\beta E_0}{\pi \rho C_p} \cdot \frac{\exp\left[-2[(x^2+y^2)/(8\alpha t + a^2)]\right]}{(8\alpha t + a^2)} \tag{10.6}$$

where a is the beam radius and E_0 is the pulse energy. A probe beam can be used to detect refractive index variations in a heated region, as discussed in Section 4.5. The refractive index changes impose the following effects on the probe beam as it travels through the heated region:

1. it will undergo a phase shift;
2. it will become defocused;
3. it will deviate from its normal path.

The phase shift of the probe beam can be detected by an interferometer. Intensity variations, caused by defocusing, can be detected by a combination of a pin-hole and photodetector. Probe beam deflections (as explained in Section 4.5) can be monitored by a bi-cell photodetector. Hence photothermal refractive index

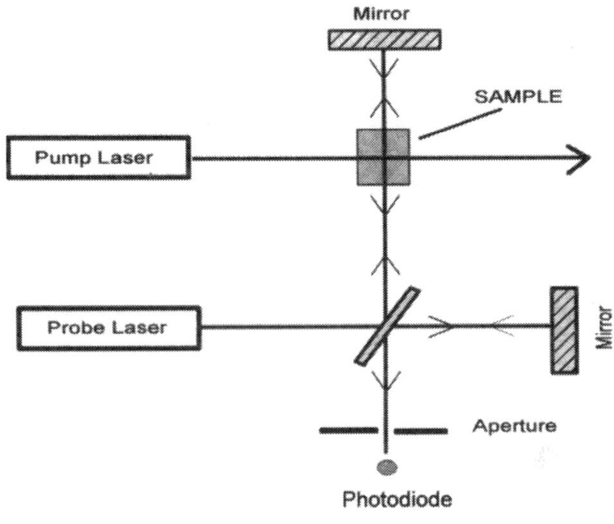

Figure 10.5 (a) *Schematic diagram of the photothermal phase shift technique.*

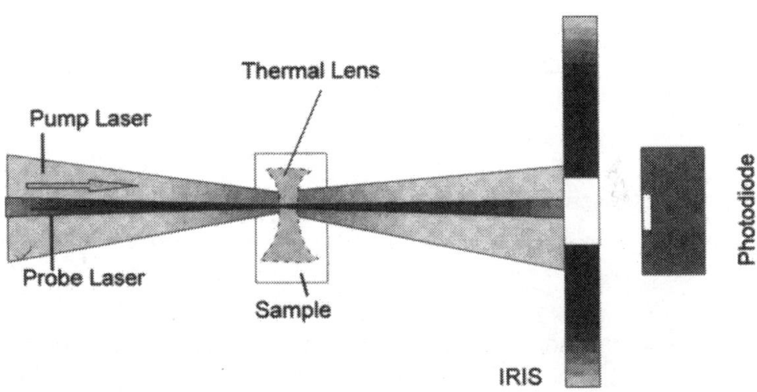

Figure 10.5 (b) *Schematic diagram of the thermal lens technique.*

modulation provides several highly sensitive, localized and non-contact detection techniques. Detection schemes based on (1) above are referred to as photothermal phase shift spectroscopy (PTPS). Those based on (2) are known as the 'thermal lens' methods, and those based on (3) are the photothermal optical beam deflection (OBD) or mirage effect techniques. Figure 10.5 illustrates these three photothermal refractive index (gradient) techniques.

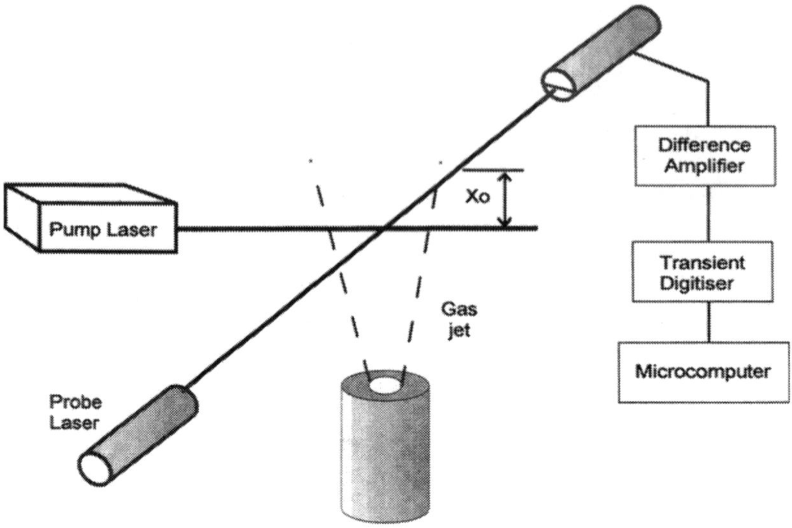

Figure 10.5 (c) *Schematic diagram of the cross beam optical beam deflection photothermal refractive index gradient technique.*

Expressions for the three types of photothermal refraction signals in a medium heated by a pulsed laser beam are:

PTPS

$$\Delta\phi(x',y',t) = \frac{4\pi}{\lambda}\int_{\text{path}} \frac{(n_0-1)T(x',y',t)}{T_A}\,dS \qquad (10.7)$$

where $\Delta\phi$ is the phase difference between the interfering waves, λ is the wavelength of the probe laser (i.e. that used in the interferometer), and n_0 and T_A are the unperturbed refractive index and temperature of the heated sample respectively.

Thermal lens

$$I(t) = I(0)\left[1-\frac{\theta}{1+t_c/t}+\frac{1}{2}\left(\frac{\theta}{1+t_c/t}\right)^2\right]^{-1} \qquad (10.8)$$

where

Table 10.2 $(dn/dT)/k$ of gases and liquids

Gases	$\times 10^{-3}(dn/dT)/k$ (cm W^{-1})	*Liquids*	(dn/dT)/k (cm W^{-1})
CO_2	–8.3	CCl_4	–0.57
Ar	–4.9	isooctane	–0.50
CH_4	–3.9	benzene	–0.44
N_2	–3.5	acetone	–0.31
O_2	–3.1	methanol	–0.19
H_2	–0.23	water	–0.013
He	–0.073		

$$\theta = \frac{1.19PA}{\lambda k}\frac{\mathrm{d}n}{\mathrm{d}T}$$

in which *PA* is the absorbed laser power and the critical time $t_c = a^2/4\alpha$.

The thermal lens signal can be increased by using a medium with high dn/dT and low thermal conductivity k. Non-polar solvents give higher signals than aqueous solutions because of their higher dn/dT and lower k values as can be seen from Table 10.2.

OBD

$$\phi_L(x,y,t) = -\frac{L}{n}\frac{\partial n}{\partial T}\frac{8\beta E_0}{\pi\rho C_p}\frac{(x-v_x t)}{(8\alpha t + a^2)^{3/2}} \times \exp\left\{\frac{-2[(x-v_x t)^2 + y^2]}{(8\alpha t + a^2)}\right\} \quad (10.9)$$

where L is the interaction path length and v_x is the fluid velocity along the direction x.

These techniques for gases and liquids have found many applications in spectroscopy, trace detection, fluid flow measurement and thermal property determination. Their advantages include: the elimination of the PA cell noise problems; all optical non-contaminating detection and the potential for remote non-contact monitoring. A few examples of some typical applications are presented below.

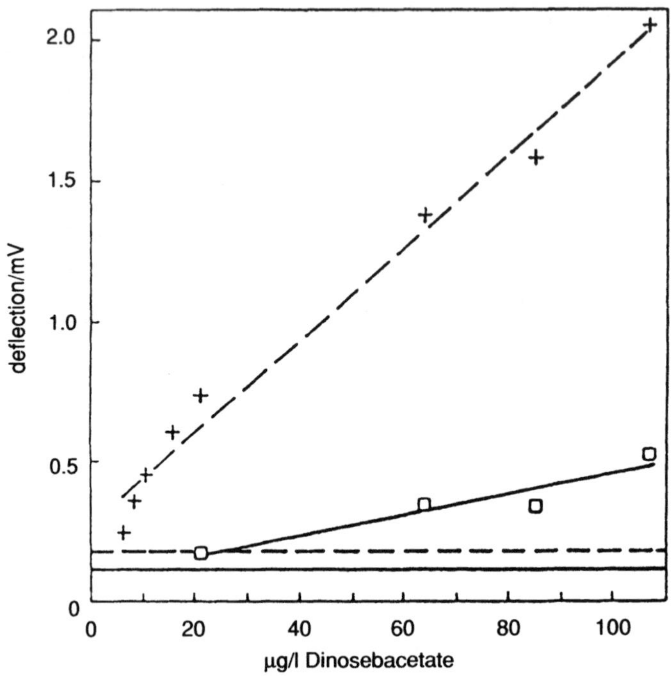

Figure 10.6 *Deflection calibration curves for the concentrations of 2-(sec-butyl) 4,6-dinitrophenolacetate in water for 1.5 m (+) and 0.25 m [□] deflection path lengths (Bohnert et al., 1992; reproduced with permission, copyright Springer-Verlag).*

10.5.1 Trace analysis

The determination of drinking water pollution by organic pesticides and metallic salts has been studied by the transverse OBD method. Figure 10.6 shows the linear variation of the OBD signal with increasing concentration of a pesticide in water; an argon ion laser pump beam and HeNe probe beam were used.

10.5.2 Thermal property measurements

The thermal transport properties of liquid and gaseous media can be determined by several photothermal techniques. Probe beam deflection methods were covered in Chapter 9. An example of an application of the thermal lens (TL) method and its sensitivity to changes in thermal parameters is shown in Fig. 10.7. The figure

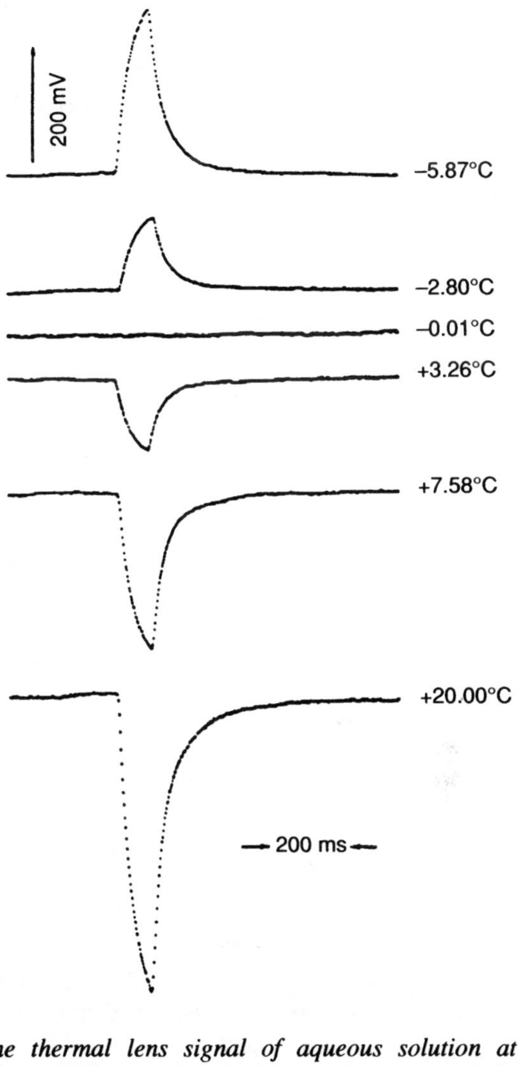

Figure 10.7 *The thermal lens signal of aqueous solution at different temperatures (Tran, 1992; reproduced with permission, copyright Springer-Verlag).*

shows the temperature variation of the TL signal from water. The change in the direction of the signal peak is due to the change in the sign of the dn/dT value of water from being positive for $T < -0.01$°C to negative for $T > -0.01$°C. At –0.01°C the TL signal becomes equal to zero (at the maximum of refractive index, dn/dT = 0). The time variation of the TL signal can be used to determine k and dn/dT.

10.5.3 Fluid flow velocity

Gas velocity measurements are important for fluid mechanics studies, aerodynamics and combustion. The main technique currently used for flow measurement is laser doppler anemometry (LDA). In this method, particles injected into the flow move through stationary fringes set up by the interference of two laser beams. The flow velocity is determined by the doppler shift of the scattered laser radiation. Photothermal methods, in particular the crossed beam probe beam deflection technique shown in Fig. 10.5(c), avoid the uncertainties introduced by the use of seed particles (particle drag, tracking, interference with combustion process). From equation (10.9) it can be seen that the pulsed heating probe beam deflection signal vanishes when $x - v_x t = 0$. Thus, knowing the probe and pump beam separation x and the time when the deflection is zero the velocity of the gas/fluid can be determined using this simple 'time-of-flight' method. Figure 10.8 presents several transverse OBD signal traces at various probe beam positions; the gas was nitrogen with a fixed flow velocity of 2.1 m s^{-1}. The traces exhibit the expected zero-crossing for all positive values of x. The zero-crossing time for x = 0.21 mm is clearly close to the value of 0.1 ms that satisfies the above relationship.

The technique offers good spatial (< 10^{-4} cm^3) and temporal resolution (< 1 ms). Its main drawback is that the gas/fluid must be absorbing to the pump laser radiation to induce heating and be transparent to the probe beam. In some cases, seeding with an absorbing medium is required to induce adequate heating (in the example shown in Fig. 10.8 the N_2 was seeded with NO_2).

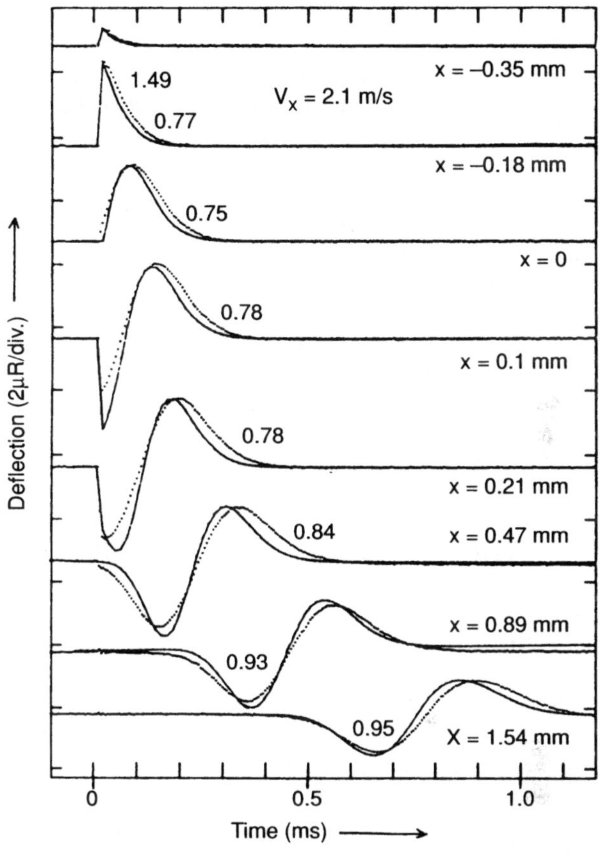

Figure 10.8 *Photothermal deflection signals in flowing nitrogen for different pump to probe separations x, and a fixed flow velocity of 2.1 m s⁻¹. The dotted and solid curves are experimental results and theoretical predictions, respectively (Rose et al., 1986; reproduced with permission).*

References

Bohnert, B., Faubel, W. and Ache, H. J., 'Remote photothermal deflection spectroscopy for trace analysis of pesticides in aqueous solutions', in *Springer Series in Optical Sciences Vol. 69: Photoacoustic and photothermal phenomena III* (ed. D. Bicanic), Springer-Verlag (1992) 46–9.

Harshbarger, W. R. and Robin, M. B., 'The opto-acoustic effect: revival of an old technique for molecular spectroscopy', *Accounts of Chemical Research*, **6** (1973).

Rose, A., Vyas, R. and Gupta, R., 'Pulsed photothermal deflection spectroscopy in a flowing medium: a quantitative investigation', *Appl. Opt.*, **25** (24) (1986) 4626–43.

Sigrist, M. W., 'Air monitoring by laser photoacoustic spectroscopy', in *Air Monitoring by Spectroscopic Techniques, Chemical Analysis Series*, **127** (ed. M. W. Sigrist), Wiley (1994) 163–238.

Tran, C. D., 'Analytical thermal lens spectroscopy: past, present and future prospects', *Springer Series in Optical Sciences Vol. 69: Photoacoustic and photothermal phenomena III* (ed. D. Bicanic) (1992) 463–473.

Further reading

Dovichi, N. J., Nolan, T. G. and Welmer, W. A., 'Theory of laser induced photothermal refraction', *Anal. Chem.*, **56** (1984) 1700–04.

Morris, M. D. and Peck, K., 'Photothermal effects in chemical analysis', *Anal. Chem.*, **58**(7) (1986) 811A–822A.

Miller, R. M., 'Industrial application of the photothermal effect', *Proc. SPIE*, **1320** (1990) 50–66.

Sigrist, M. W., 'Environmental and chemical trace gas analysis by photoacoustic methods', in *Progress in photothermal and photoacoustic science and technology Vol 1* (ed. A. Mandelis), Elsevier Science (1992).

Tam, A.C., 'Photoacoustics: Spectroscopy and other applications', *Ultrasensitive laser spectroscopy*, Academic Press (1983).

West, G. A., 'Photoacoustic spectroscopy, *Rev. Sci. Instrum.*, **54**(7) (1983) 797–817.

Index

Made in the USA
Monee, IL
13 September 2026

62978971R10142